Transistors, Diodes, and Solar Cells:

An Introduction to the Semiconductor Devices That Run the World

Shamus McNamara

Shamus McNamara
Department of Electrical and Computer Engineering
University of Louisville
Louisville, KY 40292
shamus.p.mcnamara@gmail.com

Rising Dragon Books.

McNamara, Shamus

    Transistors, Diodes, and Solar Cells: An Introduction to the Semiconductor Devices That Run the World.

    1. Engineering   2. Semiconductors

    ISBN: 978-1-960160-40-9

    10 9 8 7 6 5 4 3 2 1

DEDICATION

I dedicate this book to my parents, who taught me to think.

I dedicate this book to my brothers and sisters, cousins, aunts, uncles, and grandparents, who taught me the value of family.

I dedicate this book to my wife, who has always encouraged and supported me in everything that I do.

I dedicate this book to my children, who bring great joy and happiness to my life.

Finally, I dedicate this book to all my students and colleagues who have asked me questions that forced me to find better answers.

PREFACE

There are many books on semiconductor devices. Why would I write one?

When teaching, I found that there were some topics that were hard for students to understand, so I supplemented with my own notes. Then I wanted to make sure my students had some material that was more up to date than what is found in the existing books, so I had to write some more notes. After a few years of teaching, I found myself with Chapters 4, half of Chapter 7, and Chapters 9, 10, and 11. I also had short, 1-2 page sets of notes on other topics. With a book half-written before I had even started, it only seemed natural to complete the book. I believe this is how many textbooks are written.

I wrote the book "Operating Principles of Semiconductor Devices," which is what this book is based on. There was only a print version and limited availability. Having taught using that book for many years, I also learned how I could improve the writing to make some topics easier to understand. Thus, I wrote this book to replace it. This book is greatly expanded over the original with more examples. In addition, there are now both print and ebook versions.

In this book, I tried to make the material as clear as I could. I have tried to be somewhat detailed when deriving equations, and I have included a number of example calculations. I always include equations for both the NMOS and PMOS transistor, and I don't expect the reader to figure out how to transform the equations to go from one to another.

Solar panels have become quite prevalent, and I find that my students pay more attention and ask more questions when I cover photovoltaics than when I teach any other topic. I did not include any political or regulatory information on photovoltaics in this book because they are constantly changing, but I highly recommend that any instructor supplement with a short discussion on the societal impact of solar panels for the simple reason that the students love to hear about photovoltaics, and you will be a very popular instructor.

Chapter 10 explains the economics of making ICs because the driving force for making transistors smaller is the economics of small transistors, not the performance enhancement. The performance enhancement is important, but the rate at which transistor become smaller would have been much slower if the economics didn't favor small transistors. I find that this is a good topic for motivating students and may be taught at any time in a semester. There is only one problem at the end of the chapter, but I feel that it is a very educational problem.

This book does not introduce the BJT. I know that every other book on semiconductor devices includes a chapter on BJTs, but BJTs are rarely used anymore and I see no reason to teach it.

There is a chapter on IGBTs because these have replaced BJTs for power applications. IGBTs are not covered in most introductory books. An IGBT has a built-in BJT, but the chapter is written such that the reader doesn't need to know anything about the BJT.

I would like to acknowledge the help of my wife, Kerridwen Mangala McNamara, who has read through portions of the manuscript and provided valuable feedback, as well as helping with the formatting of the book.

Shamus McNamara
2024

Table of Contents

Nomenclature

$\blacksquare_n$	When a variable is subscripted with the letter 'n', it refers to a property of the electrons. For example, the mobility of electrons is μ_n.
$\blacksquare_p$	When a variable is subscripted with the letter 'p', it refers to a property of the holes. For example, the mobility of holes is μ_p.
$\blacksquare_N$	When a variable is subscripted with the capital letter 'N', it refers to a property in the n-region of a device. For example, the physical width of the n-region is W_N.
$\blacksquare_P$	When a variable is subscripted with the capital letter 'P', it refers to a property in the p-region of a device. For example, the physical width of the p-region is W_P.
α	This symbol serves multiple purposes as described below.

	α	Temperature Coefficient of Resistance (TCR). This parameter is used to describe the change in resistance for a change in temperature. Units: (K^{-1}) or $(°C^{-1})$.
	α	A parameter that is used for calculating yield using a binomial distribution to account for clustering of defects. This parameter is normally fit to experimental data and typical numbers are between 3 and 4.
	α	Absorption coefficient. This describes how rapidly light is absorbed and is the inverse of the absorption length. Units: (cm^{-1})

β	Internal current gain of an IGBT. The ratio of the collector current divided by the drain current due to the MOSFET built into the IGBT.
Δn	Excess electron concentration in the conduction band. $\Delta n = n - n_0$. The excess electron concentration can be positive or negative.

Δp	Excess hole concentration in the valence band. $\Delta p = p - p_0$. The excess hole concentration can be positive or negative.	
χ	Electron affinity of a semiconductor. Electron affinity is defined as $\chi = E_{vac} - E_C$. Units: (eV)	
ϵ	Dielectric constant or permittivity (two names for the same thing). This can be either a relative dielectric constant, which is unitless, or it can be the dielectric constant of free space, which has units of (F/cm).	
	ϵ_0	Permittivity of free space. $$\epsilon_0 = 8.854 \times 10^{-14} \text{ F/cm}$$
	ϵ_S	Relative dielectric constant of a semiconductor.
	ϵ_i	Relative dielectric constant of an insulator.
$\mathcal{E}$	Electric field. The electric field is technically a vector, but in this book we will only use the electric field in one dimension, making it a scalar. Units: (V/cm)	
	$\mathcal{E}_x$	One dimensional electric field in the x-direction. Units: (V/cm)
	$\mathcal{E}_S$	Critical electric field that separates the drift velocity between the non-velocity saturated regime and the velocity saturated regime.
	$\mathcal{E}_{crit}$	Electric field at which Avalanche breakdown occurs.
	$\mathcal{E}_m$	Maximum electric field in a p-n junction.
ϕ	Electric potential within a semiconductor. Units are (V).	
	ϕ_i	Electric potential of the middle of the band gap. This is related to E_i.
	ϕ_F	Electric potential of the Fermi level of the semiconductor. This is related to E_F.
	ϕ_B	In a MOSFET, the Body potential of the Fermi level relative to the intrinsic level.
	ϕ_{SB}	Potential at the surface of a MOSFET, relative to the potential deep in the substrate. In many books the subscript 'B' is implied.
	ϕ_{bi}	Built-in electric potential of a p-n junction.
Φ	Work function in terms of electric potential. Units are volts (V).	
	Φ_G	Work function of the gate.

	Φ_B	Work function of the semiconductor substrate (body).
	Φ_{GB}	Work function difference between the gate and the semiconductor substrate (body).
γ	Injection efficiency. The ratio of the hole current being injected into the base of an IGBT from the collector divided by the total current into the base.	
κ	Thermal conductivity. Units: $\left(\frac{W}{m \cdot K}\right)$	
μ	Mobility of an electron or hole. Units: $(cm^2/V \cdot s)$	
	μ_n	Electron mobility. Units: $(cm^2/V \cdot s)$
	μ_p	Hole mobility. Units: $(cm^2/V \cdot s)$
ρ	Charge density per unit volume. Units: (C/cm^3).	
ϱ	Resistivity. Units: $(\Omega \cdot cm)$ Unfortunately, both charge density and resistivity traditionally use the same symbol. To differentiate, the symbol for resistivity used in this book is a variant of the Greek letter rho.	
σ	Conductivity. Units: $\left(\frac{1}{\Omega \cdot cm}\right)$	
Θ	Flux of particles moving through a cross-sectional area per unit time. Normally this will be a flux of electrons or holes used to describe diffusion. Units: $(cm^{-2}s^{-1})$	
	Θ_n	Flux of electrons.
	Θ_p	Flux of holes.
τ	Lifetime of an electron or hole. This is the average time before the electron or hole recombines.	
A	Area. Units: (cm^2)	
	A	Commonly used for the cross-sectional area through which current flows. The area may be width x height.
	A_{Die}	Area of one die on a substrate.
B	Base transport factor. The percentage of electrons (or holes) that make it through the base of an IGBT without recombining. Unitless.	

C	Capacitance per unit area: Units: (F/cm^2)	
	C_i	Gate capacitance per unit area due to the gate insulator on a MOSFET. Units: (F/cm^2)
	C_D	Depletion capacitance per unit area. This variable can be associated with multiple types of devices, including MOSFETs and diodes. Units: (F/cm^2)
	C_G	Total gate capacitance per unit area. This includes C_i and C_D. Units: (F/cm^2)
	C_{TOT}	Total capacitance, including the area. $C_{TOT} = C \cdot A$. Units: (F)
D	This symbol does double duty, as described below.	
	D	Defect density on a substrate. Units: # defects/area, or (cm^{-2}).
	D_n	Diffusion coefficient for electrons. Units: $\left(\frac{cm^2}{V \cdot s}\right)$
	D_p	Diffusion coefficient for holes. Units: $\left(\frac{cm^2}{V \cdot s}\right)$
E	Energy. Units are electron-volts, (eV). Note that all energies are relative, similar in manner to how voltages are relative.	
	E_{12}	When more than one subscript is present, this presents the difference between two energies. The convention is $E_{12} = E_1 - E_2$.
	E_C	Energy level of the bottom of the conduction band.
	E_V	Energy level of the top of the valence band.
	E_F	Energy of the Fermi level. A quantum state with an energy equal to the Fermi level has a 50% change of being occupied by an electron.
	E_i	Energy of the middle of the band gap.
	E_G	Band gap of a semiconductor. This is equal to $E_C - E_V$. The band gap is a material property.
	E_{vac}	Vacuum level of a material. This is the energy of an electron that is not bound by an atomic orbital.
	E_{ph}	Energy of a photon.

$f(E)$	Distribution function. Often, when there is no subscript, it will refer to the Maxwell-Boltzmann distribution function.	
	$f_{FD}(E)$	Fermi-Dirac distribution function. This is the most accurate distribution function that represents how electrons are distributed as a function of energy.
	$f_{MB}(E)$	Maxwell-Boltzmann distribution function. This function is a good approximation for the Fermi-Dirac distribution function and is much easier to handle for hand calculations.
F	Force. Force is technically a vector, but in this book we will only need to use the force in one dimension. Units: (N)	
	F_x	Force in the x-direction. Units: (N)
G	Generation rate. This is the rate at which electrons and holes are generated per volume. Units: $(\text{cm}^{-3}\text{s}^{-1})$	
H	Height of a cross-sectional area. Units: (cm)	
I	The current into a semiconductor device. Units: (A)	
	I_D	Drain current in a MOSFET.
	I_{DS}	Drain current in a MOSFET. This is the same as I_D. Although the subscript 'S' isn't required, sometimes it is written this way to emphasize that the current flows from the drain into the source.
	I_G	Gate current in a MOSFET. This is normally either zero, or very close to zero.
	I_C	Collector current in an IGBT.
	I_E	Emitter current in an IGBT.
J	Current density within the semiconductor. The current density is current divided by cross-sectional area: $J = I/A$. Units are (A/cm^2)	
	J_{diff}	Diffusion current density. This does not include the drift current density, but does include both electron and hole diffusion current densities.
	J_{drift}	Drift current density. This does not include diffusion current density, but does include both electron and hole drift current densities.
	J_n	Electron current density. This does not include hole current density, but does include both diffusion and drift current densities.

	J_p	Hole current density. This does not include electron current density, but does include both diffusion and drift current densities.
	J_{TOT}	Total current density. This is often just written as J, but the subscript may be used to emphasize that components of the current density were summed up. For example: $$J_{TOT} = J_{\text{diff}} + J_{\text{drift}}$$ $$J_{TOT} = J_n + J_p$$
	J_S	Reverse saturation current density of a p-n junction. This is the amount of current that flows under reverse bias.
	J_{op}	Optical current density. This current density is due to light incident on a semiconductor p-n junction.
	J_{sc}	Short-circuit current in a p- junction illuminated with light (photovoltaic cell).
	J_{ECB}	Tunneling current density where the electrons tunnel into the conduction band.
	J_{EVB}	Tunneling current density where the electrons tunnel into the valence band.
k		Wave number of an electron. Units: (cm^{-1})
k_B		Boltzmann's constant: $1.38 \times 10^{-23} \frac{J}{K} = 8.62 \times 10^{-5} \frac{eV}{K}$
L		Length of a layer. Units: (cm)
	L	In a MOSFET, L will refer to the length of the gate; the distance from the source to the drain.
	L_{drawn}	In a MOSFET, L_{drawn} is the distance between the source and drain as indicated on a CAD program, and is normally the drawn length of the gate metal. This length may be different than the effective gate length of a MOSFET.
	L_n	Diffusion length for electrons. This is the average distance an excess electron travels before recombination.
	L_p	Diffusion length for holes. This is the average distance an excess hole travels before recombination.

	L_A	Absorption length. This is the average distance light travels in a semiconductor before it is absorbed. The absorption length depends on the wavelength of the light.
n		Concentration of electrons in the conduction band. Units are electrons per volume, or (cm^{-3}).
	n_0	Equilibrium concentration of electrons in the conduction band.
	n_i	Intrinsic (undoped) concentration of electrons in the conduction band. This variable is commonly used in equations, and the intrinsic value should be used even if the semiconductor is doped.
N		Concentration or number density. This is used for the number density of quantum states and for the number density of dopant atoms. Units are number per volume, or (cm^{-3}).
	N_A	Concentration of acceptor atoms. Units are (cm^{-3}). Sometimes this is written as N_A^- to emphasize that these atoms are negatively ionized.
	N_D	Concentration of donor atoms. Units are (cm^{-3}). Sometimes this is written as N_D^+ to emphasize that these atoms are positively ionized.
	N_C	Effective density of states for electrons in the conduction band. Units are (cm^{-3}). This is a material property.
	N_V	Effective density of states for holes in the valence band. Units are (cm^{-3}). This is a material property.
	N_{SS}	Density of surface states on the semiconductor surface for a MOSFET. Units: (cm^{-2})
	N_{TI}	Threshold adjust Implant density. This is the atomic density introduced by an ion implanter used to adjust the threshold voltage of a MOSFET.
p		Concentration of holes in the valence band. Units are holes per volume, or (cm^{-3}).

xxii   *Transistors, Diodes, and Solar Cells*

	p_0	Equilibrium concentration of holes in the valence band.
	p_i	Intrinsic (undoped) concentration of holes in the valence band. This variable is not normally used in equations because its value is the same as n_i.
P		Power. This can be electrical power or optical power. The optical power will always have a subscript. Units: (W)
	P_{ph}	Optical power. Units: (W)
q		Charge of an electron or hole. The constant q is always a positive number. The equations that use q have a positive or negative sign to represent that charge, as appropriate. $q = 1.6 \times 10^{-19}$ C.
Q		Charge density per unit area. Units: (C/cm^2)
	Q_D	Depletion charge per unit area.
	Q_S	Charge per unit area in the semiconductor.
	Q_G	Charge per unit area on the gate of a MOSFET.
	Q_i	Inversion charge per unit area in the channel region of a MOSFET.
	Q_{SS}	Charge density due to the surface states on a semiconductor surface. This charge is always positive. $Q_{SS} = qN_{SS}$
	Q_{TI}	Threshold adjust Implant charge density. This is a surface charge density introduced by an ion implanter used to adjust the threshold voltage of a MOSFET.
R		The letter R does double duty, as shown below.
	R	Resistance. Units: (Ω).
	R_N	Resistance of the n- region.
	R_P	Resistance of the p- region.
	R_{eq}	Equivalent resistance.
	R	Recombination rate. The rate at which electrons and holes recombine, per volume. Units: (cm^{-3}s^{-1})
S		The so-called sub-threshold slope for a MOSFET. This is the voltage required to obtain a 10x change in current when the MOSFET is in cutoff.

t	Time. Units: (s)	
t_i	Thickness of the insulating layer on a MOSFET. Units: (cm)	
t_{si}	Thickness of a silicon layer, especially in a SOI MOSFET or FinFET. Units: (cm)	
T	Absolute temperature. Units: (K)	
v_d	Drift velocity of an electron or hole. This is the average velocity of the carrier due to an electric field. Units: (cm/s)	
V	An externally applied voltage. Unfortunately, conventions require that subscripts are sometimes repeated, such as reusing V_B for the base of a BJT and the body of a MOSFET. This book will ensure that the subscripts are unique for any given device. Units are volts (V).	
	V_{12}	When more than one subscript is present, this presents a voltage between two points. The convention is $V_{12} = V_1 - V_2$.
	V_G	Gate voltage on a MOSFET.
	V_D	Drain voltage on a MOSFET.
	V_D	Applied voltage on a diode when modeling the diode with a series resistor. The voltage across the depletion region, V_{pn}, is smaller than the voltage applied to the electrodes, V_D.
	V_S	Source voltage on a MOSFET. In most equations the source will be the reference voltage, and the other terminal voltages will be written as V_{DS}, V_{GS}, and V_{BS}.
	V_B	Body (or substrate) voltage on a MOSFET.
	V_T	Threshold voltage for a MOSFET. This is the voltage that must be applied to the gate to turn on a MOSFET.
	V_{FB}	Flat band voltage for a MOSFET. This is the voltage that must be applied to the gate to make the energy bands flat within the semiconductor.
	V_{app}	Voltage applied to a semiconductor.
	V_{pn}	Voltage applied between the anode (p-region) and cathode (n-region) of a p-n junction.

	V_E	Emitter voltage on an IGBT.
	V_C	Collector voltage on an IGBT.
	V_{oc}	Open-circuit voltage in a p-n junction illuminated with light (photovoltaic cell).
W	Width of a layer. Units: (cm)	
	W	In a MOSFET, W refers to the width of the gate.
	W_D	Width of the depletion layer. In a p-n junction, W_D refers to the total depletion layer width of both the p- and n- regions. Units: (cm)
	W	Width of a cross-sectional area.
	W_P	Physical width of the p- region, including any depletion regions.
	W_N	Physical width of the n- region, including any depletion regions.
x	Coordinate axis. Sometimes x is subscripted to indicate specific points on the coordinate axis.	
	x_N	In a p-n junction, x_N is the width of the depletion region in the n- region. The edge of the depletion region is located at $x = x_N$.
	x_P	In a p-n junction, x_P is the width of the depletion region in the p- region. The edge of the depletion region is located at $x = -x_P$. Thus x_P is a positive number.
Y	Yield. The percent of good die on a wafer. This is unitless, but commonly expressed as a percentage.	

1 Introduction

Semiconductors are present everywhere in modern society. The reason semiconductors are so prevalent is because the electrical properties are easily modified through design and through other physical effects. In addition, making electronic devices out of semiconductors is economical.

Semiconductors are not special because they conduct electricity poorly. Many materials conduct electricity poorly. In fact, we can take a thin wire of copper and make it very long so that it will have a high resistance. The resistance can then be higher than the resistance of a short length of a semiconductor. So what, then, makes a semiconductor special? Semiconductors are special because their resistivity can be changed with light or an applied voltage on demand. A comparison is shown in the following table.

	Conductor	Semiconductor	Insulator
Example material	Copper	Silicon	Glass
Resistivity	Low	High	High
Conductivity	High	Low	Low
Resistivity change with:			
Light	None	Large	None
Voltage	None	Large	None
Temperature	Small	Large	Small

It is the large change in resistivity with voltage that will be used most often throughout this book because it permits us to create diodes, transistors, and power transistors. Solar cells will take advantage of the change in resistivity of both voltage and light.

1.1   Uses of Semiconductors

Semiconductors are used all around us. They make up our computers, cell phones, and tablet computers. They are integral to the television screens that we watch. They are used in the home appliances that we utilize, such as the microwave oven, refrigerator, oven, dishwasher, and clothes dryer. They are integral to the cars we drive, as well as controlling the traffic lights that direct traffic. They are used in LEDs that are used in lighting for computer screens, traffic lights, Christmas tree lights, and other lights. They are used in all solar panels on homeowners' rooftops, commercial buildings, and in large industrial solar plants. Semiconductors are used wherever you see a blinking light: on DVD players, MP3 players, wrist watches, fitness bands, exit signs, and more. Finally, they are integral to the Internet, from electronic interfaces such as Wi-Fi and Ethernet, to the semiconductor laser diodes and detectors that are used in fiber optic communications.

1.2   Understanding Concepts

In this book, it is very important to learn concepts. In many engineering courses, a small number of concepts are taught and you learn how to apply these concepts to a variety of situations. For example, in a Circuits Course, you learn Ohm's law, Kirchhoff's current and voltage laws, and a few equations for inductors and capacitors. From these basic laws, an entire semester's worth of RCL circuits are analyzed.

In this book many, many concepts will be introduced, and only a few applications will be shown. This is the exact opposite of most engineering courses, in which a few concepts are used to explore a large variety of applications. Thus, it will be imperative for the reader to spend a lot of time reviewing the concepts. While an attempt will be made to introduce each concept in a simple manner such that concepts build on each other, it will be found that many concepts interact with other concepts. This book will try to help with the understanding by re-introducing concepts periodically, assuming that the reader has a better understanding of the underlying concepts and will better grasp the concepts being reviewed. But the reader of the book will find it is beneficial to re-read portions of the book a second time to gain a thorough understanding of the concepts. The author has taught this subject many times over the years, and it is his experience that the concepts must be reviewed more than once to gain a good understanding.

1.3   Outline of the book

This book starts with an introduction to the physics of semiconductors, including band diagrams, doping, resistivity, and current flow. Other semiconductor physics, such as optical absorption, E-k diagrams, continuity equations, generation, and recombination are deferred until they are needed for specific semiconductor devices. The philosophy of the book is to get to the device operation as quickly as possible, and learn the physics along the way.

MOSFETs are introduced as the first device because they are so important, and because it provides a great opportunity to describe what happens when energy bands bend: depletion regions and inversion regions are formed. Using a MOSFET, a depletion region can be introduced without worrying about the concept of current flow through the depletion region.

Then diodes are introduced, which requires a good understanding of minority carriers, diffusion current, carrier recombination, and the continuity equation.

Solar cells follow diodes because solar cells are a special type of diode. The chapter on solar cells covers the physics of how semiconductors absorb light. Similarly, LEDs and lasers follow solar cells because they are special types of diodes. This chapter covers the physics of how semiconductors emit light.

Now the reader is ready for advanced MOSFET topographies. Sub 100-nm MOSFETs are described because they are commonly used in industry and are state of the art. These devices also look and behave very differently from the long-channel MOSFET introduced earlier.

The final chapter covers power transistors. The power MOSFET and IGBT are covered. It is covered to provide the reader with an introduction to power semiconductor devices, which are used to control the flow of power to other devices. For example, they are used in power supplies, USB chargers, and motor controllers.

1.4   Homework Problems

1. Consider a slab of silicon with a resistivity of $10 \ \Omega \cdot cm$. It has a length of 5 cm, width of 1 cm, and a thickness of 1 cm. Current flows along the length of the silicon.
   a. What is the resistance of the silicon?
   b. If I were to replace the silicon with copper (resistivity = $1.68 \times 10^{-6} \ \Omega \cdot cm$), what thickness gives the same resistance as you obtained in part (a)?

Note #1:   Using a technique called atomic layer deposition (ALD), it is possible to get copper just a few atoms thick. Therefore, your answer is reasonable. Look up Atomic Layer Deposition!

Note #2: The resistivity of copper increases when the thickness is much smaller than a typical grain size due to increased scattering. We did not take this into account.

2   Semiconductor Materials

This chapter is a brief outline of the uses of different types of semiconductors. The purpose of this chapter is to motivate the reader to take an interest in the subject material. The reader may not be familiar with some concepts used in this chapter, but the author hopes that this material is sufficiently general to pique the interest of the reader. After learning more about semiconductors in later chapters, the reader may wish to re-visit this chapter.

2.1   Silicon

Silicon (Si) is the most widely used semiconductor. It has been said that we have gone through the Stone Age, the Bronze Age, the Iron Age, … and that we are now in the Silicon Age. Silicon easily makes up over 99% of the semiconductor market.

Silicon is used to make transistors for analog circuits, digital circuits, and power circuits. Silicon can be grown as a single crystal with relative ease and incredible purity, and thus is available at a lower cost than any other semiconductor. Further, it is stable at high temperatures, engineers know how to introduce dopants with nanometer precision, and a stable insulator can be readily added that introduces very few defects. These advantages cannot be met by any other semiconductor. The manufacturing knowledge that has been built up over the past 50+ years creates a huge market barrier to the introduction of a new semiconductor material to replace silicon.

Silicon is used to create VLSI circuits that consist of billions of transistors on an integrated circuit. The reason silicon is used for VLSI is that the transistors can be made very small and use much less power than the transistors used with any other semiconductor. When there are 1 billion transistors on a chip, 1 μA of leakage current per transistor is equal to 1000 Amps of total wasted current.

Silicon is used for most solar cells. For solar cells, silicon comes in three forms: amorphous, polycrystalline, and single crystal. The efficiency improves as the silicon is made more crystalline, but the cost also increases. Single crystal solar cells are widely available with efficiencies over 20 %.

Silicon is used for power diodes, MOSFETs and IGBTs. Power BJTs were made of silicon, but the IGBT has largely displaced the BJT market.

Silicon is not the fastest semiconductor, but the faster semiconductors cannot compete in terms of power consumption, and the other semiconductors do not have a sufficiently large speed advantage to overcome the problems with power consumption for most applications.

If you look at your computer, cell phone, tablet, electronics in your car, electronics in your appliances, etc., you will find that it is almost all silicon based.

2.2   Germanium

Germanium (Ge) was used to create the first transistor, but it is much harder to work with germanium than silicon. The natural oxide, germanium oxide, is water soluble making manufacturing much more difficult. Germanium has a smaller band gap than silicon leading to higher leakage currents. These disadvantages led to the rapid adoption of silicon as the semiconductor of choice.

Germanium, with its small band gap, may be for infrared detectors.

2.3   SiGe

Silicon germanium (SiGe) is sometimes used on a silicon substrate because it can be readily grown on the silicon substrate with few defects. By changing the amount of germanium, the band gap can be changed and a built-in electric field can be generated. This is used to provide very fast heterojunction bipolar transistors that are faster than silicon MOSFETs. These transistors often used in high frequency RF applications, such as the RF interface on a cell phone.

2.4   GaAs

Gallium Arsenide (GaAs) has a higher electron mobility than silicon, and thus transistors made of GaAs are significantly faster. Two major disadvantages of GaAs are: (1) GaAs is much more expensive to manufacture, and (2) the power consumption is high. For a time, there was great interest in researching GaAs to replace silicon as the semiconductor of choice, but these disadvantages could never be overcome. Today, GaAs is used for high frequency RF applications, such as the RF interface on a cell phone.

GaAs is also a great light emitter, and is thus commonly used for making LEDs and laser diodes. By changing the chemical composition of GaAs, such as by adding Aluminum or Phosphorus, the wavelength of the emitted light may be modified. When you see a LED, or use a laser pointer, it is almost always made using GaAs.

2.5   GaN

Gallium Nitride (GaN) is a wide band gap semiconductor that readily emits blue light. At the time this is written, the defect density of GaN is many orders of magnitude higher than silicon, and extensive research efforts are underway to discover a method of creating inexpensive GaN substrates with lower defect densities. GaN laser diodes are used in Blu-Ray readers.

Despite the high defect density, GaN has attracted great interest for power transistors because of its high breakdown voltage and faster switching speed compared to silicon.

GaN LEDs make white LEDs possible since GaAs cannot produce blue light. White-light LEDs have become an economical solution for lighting applications as GaN substrates become less expensive.

GaN is also used for High Electron Mobility Transistors (HEMT) for high frequency RF applications.

2.6   SiC

Silicon carbide (SiC) is a wide band gap semiconductor that is used for power semiconductor devices. Having a large band gap, SiC has the same advantages as GaN for power devices: higher breakdown voltage;

faster switching speed, and lower switching losses compared to silicon. But SiC also benefits from being very similar to silicon in terms of fabrication, making it very attractive from a manufacturing perspective. SiC is more expensive than silicon, but the reduced power losses make SiC transistors the best solution for many high power applications.

2.7   Bismuth Telluride

Bismuth Telluride (Bi_3Te_4) is a small band gap semiconductor that has good thermoelectric properties. By running a current through it, one side gets hot while the other side gets cold. This property can be used to cool electronics without any moving parts. The efficiency is not as great as a refrigerator or AC unit, so that market penetration is not large, but the simplicity and small size makes it attractive for some applications.

2.8   Diamond

Diamond is a form of carbon. It is commonly considered an insulator, but some researchers are investigating its use as a wide band gap semiconductor. Diamond has a very high thermal conductivity, making it potentially useful for power electronics where it will be much easier to cool the chip. Its thermal conductivity of 2200 W/m·K is about 5x greater than copper. Theoretically, diamond can provide very fast electronics with low leakage currents. The difficulties that must be overcome include problems with the manufacture of defect free diamond and problems with doping diamond. Until a manufacturing breakthrough occurs, diamond is not likely to be widely used as a semiconductor material.

2.9   Graphene

By itself, graphene is not technically a semiconductor because it has zero band gap. However, a band gap usually arises when other materials are added to graphene. Graphene is the subject of intense research because it is a 2 dimensional material. That is, graphene is made up of a single atomic layer of carbon atoms. This creates many unique physical, electronic, optical, and magnetic properties that are under investigation.

It is unclear what the future holds for graphene. The unique electronic properties mean that it is unlikely that graphene-based transistors will look like their silicon counterparts.

In addition to graphene, there are numerous other 2D materials that are under investigation, many of which have a natural band gap. This is a ripe area for research.

2.10 Organic Semiconductors

There is great interest in making semiconductors out of organic materials instead of inorganic materials. Organic materials are typically less expensive and processed at lower temperatures. However, the organic materials developed thus far have much lower electron mobility and hole mobility than silicon, and thus are much slower and cannot be used for power devices. Organic semiconductors have started to be commercialized, such as for OLED displays.

2.11 Problems

1. Do some investigating and find a typical cost for a substrate of the following semiconductors: Si, Ge, GaAs, and GaN.

    a. Make a table with the following five columns: semiconductor type (e.g., Si), wafer diameter, wafer cost, cost per cm^2, and your reference.

       Note: You should find that silicon is the least expensive per area.

    b. Explain why silicon is a popular semiconductor for manufacturing.

3 Energy Bands and Current

This book starts by examining how current flows in a semiconductor. We will find that Ohm's law is not sufficient to describe the current flow, and the explanation will require an understanding of how the electrons are arranged into atomic orbitals. We will also discover how we can modify the resistivity of a semiconductor. By the end of this chapter, we will have a thorough understanding of the difference between conductors, semiconductors, and insulators.

3.1 Chemical Bonding and Atomic Orbitals

All atoms are made of a positively charged nucleus with a cloud of electrons surrounding the nucleus. Sometimes the electrons are illustrated as a group of electrons orbiting the nucleus, similar to how planets orbit a star. However, this is a poor model, as the electrons are better represented as waves that occupy a large amount of space around, and possibly within, the nucleus. The "orbits" that are permitted around the nucleus are called **atomic orbitals**. From chemistry class, the atomic orbitals include: 1s, 2s, 2p, 3s, 3p, 3d, ... Two electrons can occupy the 1s orbital due to spin. The p- orbitals are really three different orbitals, and they are often designated as p_x, p_y, and p_z. A **quantum state** is a configuration that an electron could be in. Due to spin, each atomic orbital has two quantum states. For example, the 1s atomic orbital has 2 quantum states, the 2s atomic orbital has 2 quantum states, and the 2p atomic orbital has 6 quantum states: $p_x \uparrow$, $p_x \downarrow$, $p_y \uparrow$, $p_y \downarrow$, $p_z \uparrow$, and $p_z \downarrow$.

In hydrogen, there is one electron and it occupies the 1s atomic orbital. In helium, the two electrons occupy the same 1s atomic orbital, but they occupy different quantum states because one electron has spin up and one electron has spin down. Both quantum states have the same energy: the energy of the 1s atomic orbital. Both electrons in these quantum states also occupy the same position: the 1s atomic orbital. Both hydrogen and helium have a 2s atomic orbital, but there are not enough

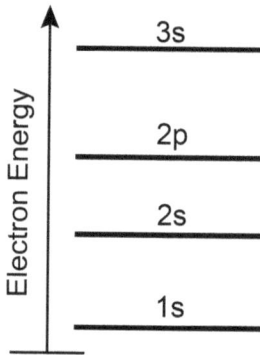

Figure 3.1: Different atomic orbitals have different energies. The spacing depends upon the element, and sometimes two atomic orbitals will have the same energy. Electrons must be in an atomic orbital, and cannot have an energy between atomic orbitals.

electrons in these elements to occupy this atomic orbital. Hence, in Hydrogen and Helium the 2s atomic orbital represents two empty quantum states. We can plot the energy of the atomic orbitals, as per Figure 3.1. Note that this figure does not indicate the number of quantum states at each energy, nor does this figure indicate how many electrons are in each atomic orbital.

A couple of important things to remember from chemistry class are:

1. Only one electron may occupy each quantum state. (Pauli exclusion principle)
2. More than one quantum state may have the same energy.
3. More than one electron may occupy the same position.
4. Covalently bonded chemicals prefer to have full outer (valence) atomic orbitals.

As individual atoms, the atomic orbitals are straightforward. When the atoms chemically bond, the atomic orbitals interact. Carbon (diamond) and silicon each contain four electrons in the topmost filled atomic orbital. For carbon, the electrons fill the $1s^2$, $2s^1$, $2p^3$ atomic orbitals. For silicon, the electrons fill the $1s^2$, $2s^2$, $2p^6$, $3s^1$, $3p^3$ atomic orbitals. The superscripts indicate the number of electrons in each atomic orbital. Note that in carbon and silicon, the uppermost 's' shell is only half filled, and the 'p' shell is half filled. This is because the two atomic orbitals have approximately the same energy. This is commonly called a sp^3 atomic orbital.

Consider what happens as two silicon atoms bond to each other. On the left side of Figure 3.2 is an isolated silicon atom. Counting spin, the sp^3

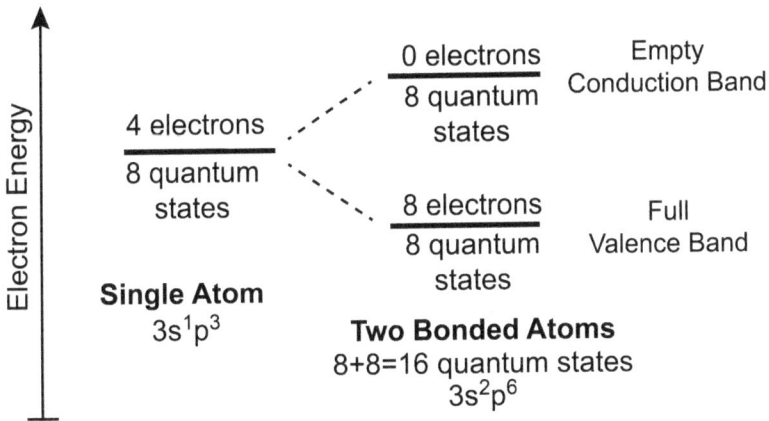

Figure 3.2: This figure schematically represents how the atomic orbitals split in energy when two atoms come together.

atomic orbital gives us eight quantum states. The isolated silicon atom has four electrons occupying the eight quantum states. When the atoms come together, the atomic orbitals of each atom interact and the energy splits. Half the atomic orbitals gain energy, and half the atomic orbitals lose energy. There are now 16 quantum states; 8 from each atom. These 16 quantum states were split in energy; 8 went to a higher energy level and 8 went to a lower energy level. There are 8 electrons; four from each atom. These 8 electrons will occupy the 8 quantum states at the lower energy, and zero electrons will occupy the quantum states at the higher energy.

Chemists call the 8 quantum states at the lower energy the bonding orbital. We will call this the **valence band**. Chemists call the 8 quantum states at a higher energy the anti-bonding orbital. We will call this the **conduction band**.

When a third atom is included, the valence band splits again, and the conduction band also splits again. The name changes from orbital, when dealing with one or two atoms, to band, when we are dealing with tens, hundreds, trillions, or more atoms. A large number of atoms requires that the atomic orbitals split many times. However, the additional splits in energy are much smaller than the first split and the resulting energy levels are clustered into two "bands".

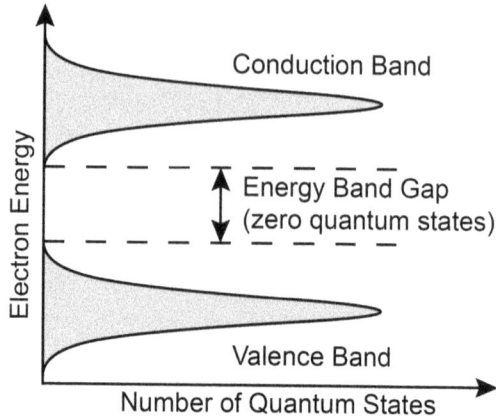

Figure 3.3: The number of quantum states (x- axis) depends upon the energy of the quantum state (y- axis). Between the conduction band and valence band is a region called the energy band gap where the number of quantum states is zero.

Figure 3.3 shows the distribution of quantum states as a function of energy for a large number of atoms. This figure shows the axis for energy in the y- direction, and the number of the quantum states in the x-direction. There are three very important concepts here:

1. There is a distribution of quantum states that are clustered tightly together called the **valence band**. This band is full of electrons, similar to the case for just two atoms described above.
2. There is a distribution of quantum states that are clustered tightly together called the **conduction band**. This band is devoid of electrons, but we could put an electron in this band.
3. There is a band of energies for which there are zero quantum states. This band of energies is called the **band gap**. An electron may not have an energy within the band gap because there are no quantum states available for the electron to occupy.

3.2 Energy Bands, Conductors, Semiconductors, and Insulators

Consider the valence band. There are four quantum states available from each silicon atom. Silicon also has four valence electrons. Thus, one electron occupies each quantum state in the valence band. In reality, a few electrons have enough energy to leave the valence band and occupy the conduction band. In silicon at room temperature, only 1 in 10^{13} electrons leave the valence band. Hence, we will say that the valence band is **nearly full**.

Now consider the conduction band. In silicon, the four valence electrons of a Silicon atom occupy the valence band, with no electrons left over. Thus, at first glance it would appear that no electrons occupy the conduction band. However, a few electrons from the valence band have sufficient energy to occupy a quantum state within the conduction band. In silicon at room temperature, only 1 in 10^{13} quantum states in the conduction band has an electron. Hence, we will say that the conduction band is **nearly empty**.

The numbers will change for other semiconductors. Whether we find the number to be 1 in 10^{10} or 1 in 10^{30}, we will still arrive at the conclusion that the valence band is nearly full of electrons and the conduction band is nearly empty of electrons.

Two or more electrons may occupy the same space. In fact, in every atom from He on up, there is a $1s^2$ shell with two electrons. One has spin up; one has spin down. Both occupy the same space: the 1s atomic orbital. This is allowed because they have different quantum states due to their differing spin.

Let us examine the electrons in the conduction band, as shown in Figure 3.4. This figure shows a few electrons in the conduction band with numerous quantum states to enter. That is, if an electric field were applied, the electrons would be accelerated because there is a quantum state available to occupy that represents the larger velocity. The more electrons there are, the more current that may flow.

The situation in the valence band is very different. If an electric field were applied, there is space for only one electron to move. The other electrons feel the force of the electric field, but they cannot move because there is no quantum state available to occupy. Thus, the electrons in the valence band provide a current that is proportional to the number of empty quantum states, not the number of electrons. Figure 3.4b shows the concept of a hole in the valence band.

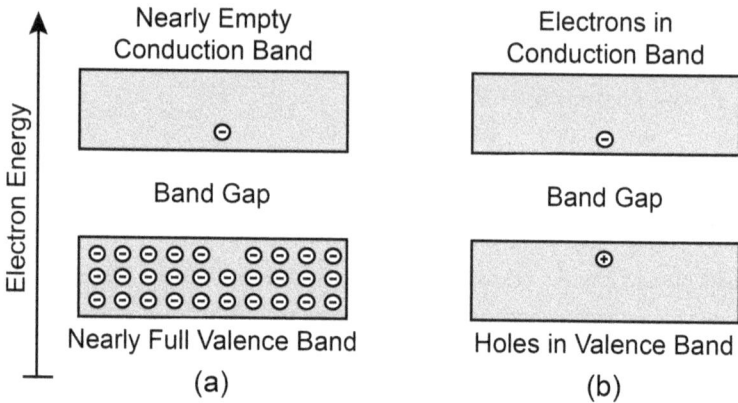

Figure 3.4: The conduction band and valence band. (a) The valence band is nearly full of electrons, but one electron is in the conduction band. (b) The valence band is shown with one hole, representing the empty quantum state left behind when the electron moved to the conduction band.

Due to the complex interactions between electrons, nuclei, and quantum mechanics, the electrons do not move as expected[1]. We define a hole as an empty quantum state surrounded by quantum states filled with electrons. Thus, an empty quantum state in the valence band, and only in the valence band, is a **hole**. The conduction band cannot contain holes because the conduction band is nearly empty of electrons, and thus there are no empty quantum states surrounded by electrons. When the electric field causes an electron in the valence band to move to the right, the hole is moving to the left. Thus, the electron moves in the opposite direction of the electric field, but the hole moves in the same direction as the electric field, as illustrated in Figure 3.5. Thus, we will say that the hole has a positive charge, and we will compute the current due to the holes. In summary, an electric field causes current flow in the valence band due to the holes.

You may have numerous questions, such as: If it is really the electrons moving, why don't holes have a negative charge? And if holes aren't real particles, how can they carry current? And

---

[1] There is more to the story. This is similar to learning about subtraction and being told, "you must subtract the small number from the big number." But there was a more advanced concept to learn: negative numbers!

Electric Field, $\mathcal{E}(x)$ Electric Field, $\mathcal{E}(x)$

t=0 ⊝⊝⊝⊝⊝ ⊝⊝⊝⊝   _ _ _ _ _ ⊕ _ _ _ _

t=1 ⊝⊝⊝⊝⊝⊝ ⊝⊝⊝   _ _ _ _ _ _ ⊕ _ _ _

t=2 ⊝⊝⊝⊝⊝⊝⊝ ⊝⊝   _ _ _ _ _ _ _ ⊕ _ _

t=3 ⊝⊝⊝⊝⊝⊝⊝⊝ ⊝   _ _ _ _ _ _ _ _ ⊕ _

Electron Representation   **Hole Representation**
Electrons move to the left   Holes move to the right in a
in a nearly full valence band   nearly empty valence band

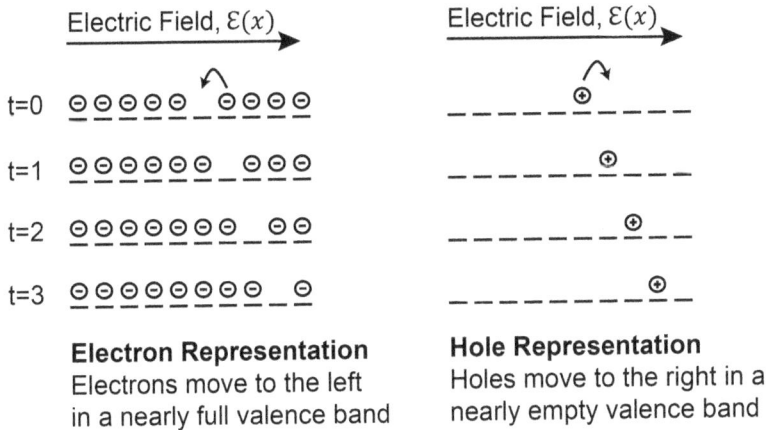

Figure 3.5: Schematic of the movement of electrons in the valence band. On the left is an electron representation, where the electrons move in the opposite direction of the electric field because they are negatively charged. On the right is a hole representation, where the holes move in the same direction as the electric field, acting as if they are positively charged.

since electrons and holes are moving, don't the negative and positive charges cancel? These are all excellent questions, and require a complicated answer with the help of quantum mechanics. The answers are available in a more advanced textbook and taught in a more advanced course. In this book I am asking you accept the concept of a positively charged hole in the valence band.

The concept of a hole is one of the most important concepts to understand. Conductors do not have holes. Semiconductors do. Read this section several times and ask for help if necessary, but you must come to believe the following key points:

1. A hole has a positive charge.
2. Holes carry the current in the valence band; electrons do not.
3. Holes only exist in the valence band; holes cannot exist in the conduction band.
4. A hole has higher energy as it moves down on the energy scale used for electrons. Or written the other way, a hole has lower energy as it moves up the energy scale for electrons.

3.3 Effective Density of States

We have seen that the density of quantum states varies with energy, and that there are zero quantum states within the band gap. The electrons in the conduction band are few and settle into the lowest energy for electrons, which is the bottom of the conduction band. The energy of the bottom of the conduction band is labeled E_C. The holes in the valence band are few and settle into the lowest energy for holes, which is the top of the valence band. The energy of the top of the valence band is labeled E_V. Therefore, these two energy levels, E_C and E_V, are very important energy levels in a semiconductor as these are generally the energies of the electrons and holes.

Although the density of quantum states varies with energy, an effective number can be found that does not depend on energy. This number is known as the **effective density of states**, and has units of # of states per volume. Since a number is unitless, the units work out to be cm^{-3}. There is an effective density of states for the conduction band, N_C, and an effective density of states for the valence band, N_V. Table 3.1 lists the effective density of states for some semiconductors.

3.4 Energy Distribution Functions

Not all the electrons in the conduction band have the same energy. Similarly, not all the holes in the valence band have the same energy. The electrons and holes have a distribution of energies. This is similar

Table 3.1: Effective Density of States for some common semiconductors.

	Conduction band effective density of states, N_C (cm^{-3})	Valence band effective density of states, N_V (cm^{-3})
Silicon	3.51×10^{19}	1.87×10^{19}
Germanium	1.02×10^{19}	5.64×10^{18}
GaAs	4.35×10^{17}	7.57×10^{18}
GaN	2.30×10^{18}	1.80×10^{19}
SiC (4H)	1.23×10^{19}	4.58×10^{18}

to gas molecules in a room. Not all gas molecules have the same energy. Some gas molecules are moving slowly, and some are moving rapidly. At atmospheric pressure, the gas molecules move approximately 60 nm before they hit another gas molecule. As a result of that collision, they may exchange energy and the molecule may increase or decrease in speed. Just as it is impractical to keep track of every gas molecule, it is impractical to keep track of every electron and hole. They will move around, collide, and individual electrons and holes will increase and decrease in energy. However, the total energy present will be fixed, and is determined by the temperature.

Since we can't keep track of every individual electron and hole, we will have to understand their behavior using statistics. There is an equation that describes the distribution of energies that electrons have, called the **Fermi-Dirac Distribution Equation**. The Fermi-Dirac distribution describes the probability of a quantum state being occupied by an electron as a function of the energy of the quantum state. The Fermi-Dirac distribution equation is:

$$f_{FD}(E) = \frac{1}{1 + \exp\left(\frac{E - E_F}{k_B T}\right)} \qquad (3.1)$$

where E_F is called the **Fermi energy**, k_B is Boltzmann's constant, and T is the absolute temperature. Before defining the Fermi energy, let us consider three energies:

1. E is much more negative then E_F: The exponent has a large negative number that evaluates to a very small number. The overall denominator is one plus a very small number, or approximately 1. And hence f_{FD} is approximately 1. Physically, this means that all the quantum states that are much more negative than the Fermi energy are filled. The $1s^2$, $2s^2$, etc. quantum states are all filled. This situation describes the nearly full valence band.

2. E is much more positive than E_F: The exponent has a large positive number, which evaluates to a very large number. In the denominator, one plus a large number is still a large number, and one over a large number is a very small number. Hence, f_{FD} is

approximately 0. Physically, this means that all the quantum states that are much more positive than the Fermi energy are nearly empty. This situation describes the nearly empty conduction band.

3. $E = E_F$: The inside of the exponent is zero. $\exp(0) = 1$, and thus f_{FD} is exactly 1/2. Physically, this means that half the quantum states with an energy equal to E_F contain an electron. In a semiconductor, the Fermi energy will almost always lie between the conduction band and the valence band, and thus lie within the band gap.

The **Fermi energy** is defined as the energy at which half the quantum states are filled. The Fermi energy is not a material property. It can be changed by doping the semiconductor (to be discussed shortly) and it can be changed by applying a voltage to the semiconductor.

An example plot of the Fermi-Dirac distribution function is shown in Figure 3.6. Here, the Fermi level is in the middle of the band gap. In this graph, the energy is on the x-axis. The maximum value of f_{FD} is 1, representing 100% of the quantum states are filled with electrons. The minimum value is zero, representing that those quantum states have zero electrons. In the valence band, the Fermi-Dirac distribution has a value very near 1, but not exactly 1. Thus, the valence band is nearly full, but has a few holes in it. In the conduction band, the Fermi-Dirac distribution has a value very near 0, but not exactly 0. Thus, the conduction band is nearly empty, but has a few electrons in it.

Another commonly used distribution function is the **Maxwell-Boltzmann Distribution Function**:

$$\boxed{f_{MB}(E) = \frac{1}{\exp\left(\frac{E - E_F}{k_B T}\right)}} \qquad (3.2)$$

The Maxwell-Boltzmann distribution function is a good approximation to the Fermi-Dirac distribution function when the energy E is at least $2k_B T$ greater than E_F. At room temperature, this condition is satisfied for most semiconductors. This function is easier to work with than the Fermi-Dirac distribution function, and hence we will usually use this equation.

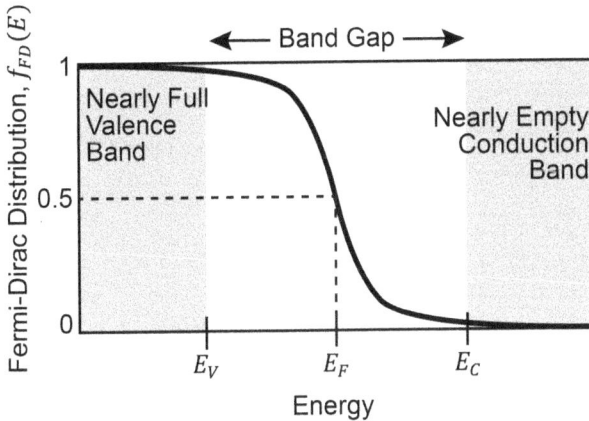

Figure 3.6: Fermi-Distribution function versus energy. Energies below E_V represent the valence band. Energies above E_C represent the conduction band.

3.5 Electron and Hole Concentrations of Pure (undoped) Semiconductors

Let's take the concepts of the effective density of quantum states, and the concept of the Fermi-Dirac distribution function. We can find the concentration of electrons in the conduction band by multiplying these together:

$$\frac{\text{\#quantum states in conduction band}}{\text{volume}} \times \frac{\text{probability of electron}}{\text{in quantum state}} = \frac{\text{\#electrons}}{\text{volume}}$$

The number of electrons per volume is known as the **electron concentration** in the conduction band with units of cm^{-3}. Usually, we will just say <u>electron concentration</u>, with the assumption that we are referring to electrons in the conduction band. In mathematical form, the electron concentration is:

$$n_0 = N_C f_{FD}(E_C) \tag{3.3}$$

Similarly for the valence band, we will find the **hole concentration** in the valence band by finding the probability that those quantum states do not have an electron:

$$p_0 = N_V[1 - f_{FD}(E_V)] \tag{3.4}$$

Let us simplify these equations by substituting the Maxwell-Boltzmann equation in place of the Fermi-Dirac equation, and inserting into the above equations. After some algebra we obtain:

$$n_0 = N_C \exp\left(-\frac{E_C - E_F}{k_B T}\right) \tag{3.5}$$

$$p_0 = N_V \exp\left(-\frac{E_F - E_V}{k_B T}\right) \tag{3.6}$$

Hence, if we know where the Fermi level is, we can calculate the electron and hole concentrations. The two equations above work for both undoped and doped semiconductors. The subscript '0' is to indicate that the electron and hole concentrations calculated using these equations are equilibrium values – that is, there is no voltage applied.

When a semiconductor is pure, it is called undoped or intrinsic. The convention is to use the subscript 'i' to indicate a variable used for an intrinsic, or pure, semiconductor in equilibrium. In a pure semiconductor, the electron in the conduction band must have come from the valence band, leaving a hole behind. Hence, the electron concentration must equal the hole concentration:

$$n_i = p_i \tag{3.7}$$

We will call this the **intrinsic carrier concentration**, without regard to the carrier type, and generally only use n_i, and not write p_i, since they have the same numeric value. It will turn out that the variable n_i is used in a lot of equations in this book, even when the semiconductor is not pure.

A very useful result arises if we multiply the electron and hole concentrations together:

$$n_0 p_0 = N_C N_V \exp\left(-\frac{E_F - E_V}{k_B T}\right) \exp\left(-\frac{E_C - E_F}{k_B T}\right) \tag{3.8}$$

Combining exponents:

$$n_0 p_0 = N_C N_V \exp\left(-\frac{E_F - E_V + E_C - E_F}{k_B T}\right) = N_C N_V \exp\left(-\frac{E_G}{k_B T}\right)$$

$$\tag{3.9}$$

Table 3.2: Intrinsic carrier concentration of some common semiconductors at room temperature.

	Silicon	Germanium	GaAs	GaN	SiC (4H)
Intrinsic carrier concentration, n_i (cm^{-3})	10^{10}	1.79×10^{13}	1.77×10^6	1.77×10^{-10}	3.10×10^{-9}

where we defined the band gap as $E_G = E_C - E_V$. We can then find n_i from:

$$n_i = \sqrt{n_i^2} = \sqrt{n_0 p_0} = \sqrt{N_C N_V \exp\left(-\frac{E_G}{k_B T}\right)} \qquad (3.10)$$

This is a very useful result. It shows that n_i depends only on temperature and material properties: N_C, N_V, E_G. Hence, the intrinsic electron concentration can also be considered a material property. The intrinsic carrier concentration of some semiconductors at room temperature are shown in the following table:

In Table 3.2, notice that the intrinsic carrier concentration is less than 1 cm^{-3} for two of the semiconductors. That is to say, there is less than one electron in the conduction band per cm^3 for a pure GaN or pure SiC semiconductor. There isn't a piece of an electron in the conduction band; electrons must be whole. Rather, the carrier concentrations are statistical values. Most of the time there are zero electrons in the conduction band, but every so often, for a short amount of time, there is an electron in the conduction band. This is a good reminder that we cannot keep track of individual electrons, and that the electrons are constantly moving, colliding, and otherwise constantly changing position and energy.

One more great result we found, but didn't have a chance to highlight, is:

$$\boxed{n_0 p_0 = n_i^2} \qquad (3.11)$$

This equation is in two boxes. Yes, two boxes. Not just one box, but two. Why? It is that important. If you can find the electron concentration, you can readily find the hole concentration from this equation, or vice versa.

For an undoped semiconductor, the number of electrons must equal the number of holes, and the Fermi level will be in the middle of the band gap. This energy level, the middle of the band gap, is labeled E_i. This will be a useful energy level to work with. Using the equations presented earlier in this section, and doing some algebra, the electron and hole concentrations may be found to be:

$$\boxed{n_0 = n_i \exp\left(\frac{E_F - E_i}{k_B T}\right)} \qquad (3.12)$$

$$\boxed{p_0 = n_i \exp\left(-\frac{E_F - E_i}{k_B T}\right)} \qquad (3.13)$$

These two equations are in boxes because they are commonly used to find the electron and hole concentrations when the Fermi level is known relative to the middle of the band gap.

3.6   Doping

The electrical conductivity of semiconductors can be changed by the addition of atoms, similar to the way that the mechanical properties of steel can be modified by adding carbon, vanadium, and other atoms.

Atoms inserted into the semiconductor to change the electrical properties of semiconductors are called **dopant atoms**. There are two types of dopant atoms: **donor atoms** and **acceptor atoms**. The concentration of these dopant atoms will be sufficiently small that the chemical composition of the semiconductor may be considered unchanged and only the electrical properties change.

A **donor atom** typically has one more valence electron than the atom it replaces. Example #1: Phosphorus is in column V of the periodic table, and so it can replace a column IV element such as Silicon or Germanium. Example #2: Oxygen can replace a nitrogen atom in GaN. Example #3: Silicon can replace a gallium atom in GaN. Donor atoms get their name because after all the atomic orbitals up to the valence band are filled, there will still be one additional electron for each donor atom. It is said that this extra electron is *donated* to conduction band of the semiconductor[2].

An **acceptor atom** typically has one fewer valence electron that the atom it replaces. Example #1: Boron is in column III of the period table, so it can replace a column IV element such as silicon or germanium. Example #2: Zinc can replace a Gallium atom in GaAs. Example #3: Silicon can replace a nitrogen atom in GaN. Acceptor atoms get their name because not all the atomic orbitals up to the valence band are filled, as there is one fewer valence electron than desired for each acceptor atom. It is said that the acceptor atom *accepts* one electron from the conduction band of the semiconductor to complete the valence band.

Notice that silicon can act as either a donor atom or an acceptor atom in GaN, depending on which atom it replaces. This was shown in both examples #3 in the paragraphs above.

The dopants will be specified as a number per volume, or concentration, similar to the definition of an electron concentration or hole concentration. The symbols used are N_A for **acceptor concentration** and N_D for **donor concentration**, with typical units of cm^{-3}. In practice, the smallest concentration for the dopant atoms is 10^{13} cm^{-3}, and the largest concentration is 10^{20} cm^{-3}.

We cannot keep track of individual electrons. We might say that a donor atom donates an electron to the conduction band, but we have no way of knowing that the electron didn't just go fill in a hole in the valence band located nearby. The electrons and holes will change concentrations as appropriate, regardless of their source.

A great approximation is that, if the semiconductor is doped with donor atoms, it will have an electron concentration equal to the dopant concentration. That is,

---

[2] The actual physics of donor and acceptor atoms is more complicated, but we need not worry about it in this book because we will only use dopants that work perfectly as described.

$$n_0 \cong N_D \tag{3.14}$$

Similarly, if a semiconductor is doped with acceptor atoms, it will have a hole concentration equal to the dopant concentration. That is,

$$p_0 \cong N_A \tag{3.15}$$

If the semiconductor has both types of dopant atoms, then the dopant with the higher concentration dominates. Given the electron concentration, or given the hole concentration, the other carrier concentration can be found from Equation 3.11, repeated here because it is so important:

$$n_0 p_0 = n_i^2 \tag{3.11}$$

Doping a semiconductor does not introduce a net charge. Doping a semiconductor with donor atoms will increase the number of electrons *in the conduction band*, but that is only changing the distribution of the electrons among the energy bands. The total number of electrons will equal the total number of protons, and the net charge will be zero. Introducing donor atoms does not make the semiconductor negatively charged. Similarly, introducing acceptor atoms does not make the semiconductor positively charged, despite the fact that the hole concentration increases. The total number of positive charge and negative charges will be equal.

The following example shows that dopant atoms increase one carrier type and decrease the other carrier type. In fact, it is normal for the carrier concentrations to change by many orders of magnitude compared to an undoped semiconductor. This ability to change the carrier concentration, and hence the resistivity, of the semiconductor is a very important feature that is used to create a variety of semiconductor devices. It is this ability to control the carrier concentrations that permits us to make devices such as the MOSFET, diode, solar cell, and IGBT.

	Example 3.1	**Example 3.2**
Semiconductor	Silicon	Silicon
Dopant Atom	Boron	Phosphorus
Dopant Type	Acceptor	Donor
Dopant Concentration	$N_A = 10^{15}$ cm^{-3}	$N_D = 10^{18}$ cm^{-3}
Approximation	$p_0 = 10^{15}$ cm^{-3}	$n_0 = 10^{18}$ cm^{-3}
	$n_0 = \dfrac{n_i^2}{p_0}$ $= \dfrac{10^{20} \text{ cm}^{-6}}{10^{15} \text{ cm}^{-3}}$ $n_0 = 10^5$ cm^{-3}	$p_0 = \dfrac{n_i^2}{n_0}$ $= \dfrac{10^{20} \text{ cm}^{-6}}{10^{18} \text{ cm}^{-3}}$ $p_0 = 10^2$ cm^{-3}
Semiconductor type	p-type	n-type
Majority Carrier	Holes	Electrons
Minority Carrier	Electrons	Holes

In these examples, several new definitions were used. The **semiconductor type** refers to the carrier type that is largest. If the electron concentration is larger than the hole concentration, it is called an **n-type semiconductor**. If the hole concentration is larger than the electron concentration, it is called a **p-type semiconductor**. In example 3.1, the hole concentration is 10^{10} times larger than the electron concentration, and thus it is called a p-type semiconductor. The **majority carrier** is the carrier type that is largest, and the **minority carrier** is the carrier type that is smallest. The use of these terms will be used throughout this book so that we can refer to semiconductors that are n-type or p-type, in which either electrons or holes predominately carry the current, without having to make every statement twice.

Example 3.3: Consider silicon doped with boron at a doping concentration of $N_A = 10^{16}$ cm^{-3}. What are the electron concentration, hole concentration, Fermi level relative to the conduction band, and Fermi level relative to the valence band? Accurately draw the band diagram and indicate E_C, E_V, E_F, and their relative distance to each other.

Answer: Boron atoms are acceptor atoms. Therefore, the hole concentration is:

$$p_0 = 10^{16} \text{ cm}^{-3}$$

The electron concentration is found using $n_0 p_0 = n_i^2$.

$$n_0 = \frac{n_i^2}{p_0} = \frac{10^{20} \text{ cm}^{-6}}{10^{16} \text{ cm}^{-3}} = 10^4 \text{ cm}^{-3}$$

where n_i for silicon was found in Appendix A. The Fermi level is found from the equations for either n_0 or p_0. Let's use the equation for n_0. Solving for E_{CF}:

$$E_{CF} = -kT \ln \frac{n_0}{N_C}$$

$$= -0.0259 \text{ eV} \ln \frac{10^4 \text{ cm}^{-3}}{3.51 \times 10^{19} \text{ cm}^{-3}} = 0.927 \text{ eV}$$

Now let's find E_{FV} from p_0:

$$E_{FV} = -kT \ln \frac{p_0}{N_V}$$

$$= -0.0259 \text{ eV} \ln \frac{10^{16} \text{ cm}^{-3}}{1.87 \times 10^{19} \text{ cm}^{-3}} = 0.195 \text{ eV}$$

Now let's draw the band diagram using these calculated values. Notice that the total distance from E_C to E_V should equal the band gap.

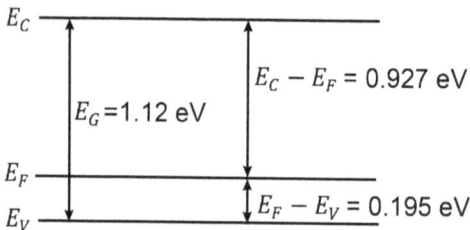

E_C

$E_C - E_F = 0.927$ eV

$E_G = 1.12$ eV

E_F

E_V

$E_F - E_V = 0.195$ eV

Time out! What were those units of 'eV' in the last example!? The symbol eV stands for electron-volt, which is a unit of **energy**.

$$1 \text{ eV} = 1.6 \times 10^{-19} \text{ Joule}$$

Because electrons are so tiny, the amount of energy that an individual electron contains is also very tiny. We have the option of either writing our answers in terms of teeny tiny values of Joules, or we can write our answers in eV. Both are correct, but I will use eV in this book.

3.7 Energy Band Diagrams

Example 3.3 is the first example of an energy band diagram as they are conventionally drawn. The x-axis is distance, and the y-axis is energy. One line is drawn to represent each of: E_C, E_V, E_i, and E_F. The energy difference between th e bottom of the conduction band, E_C, and the top of the valence band, E_V, is always fixed and equal to the band gap, E_G. The band gap for many materials may be found in Appendix A. The middle of the band gap, E_i, is exactly at the midpoint between E_C and E_V. The left side of Figure 3.7 shows a semiconductor with E_i in the middle of the band gap. I like to draw E_i with a dotted line to make it easier to spot, but this is not required.

The Fermi level always lies within the band gap of the semiconductor. Looking at Equation 3.12, the term inside the exp() term is positive when E_F is more positive than E_i, making the electron concentration large. That is, the semiconductor is n-type (majority carrier electrons) when the Fermi level is in the top half of the band gap. Figure 3.7 (right) shows an n-type semiconductor.

Looking at Equation 3.13, the term inside the exp() term is positive when E_F is more negative than E_i, making the hole concentration large.

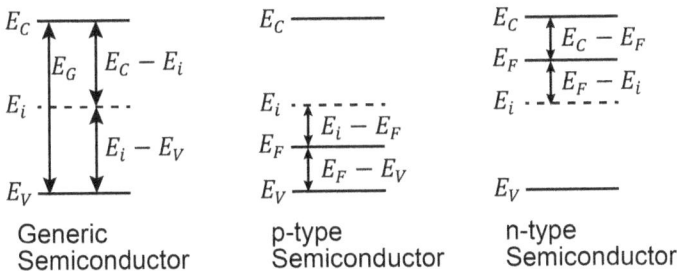

Generic Semiconductor

p-type Semiconductor

n-type Semiconductor

Figure 3.7: Energy band diagram of a semiconductor. (left) E_i is always in the middle of the band gap. (middle) E_F is always below E_i in a p-type semiconductor. (right) E_F is always above E_i in an n-type semiconductor.

That is, the semiconductor is p-type (majority carrier holes) when the Fermi level is in the bottom half of the band gap. The middle diagram of Figure 3.7 shows a p-type semiconductor.

3.8   Current Density

We will use current when talking about the current entering or leaving a semiconductor, but we will use current density when discussing current within a semiconductor. This permits us to write equations without having to be concerned about the dimensions of the semiconductor. Current density is the current divided by the cross-sectional area through which the current flows. Current density has units of A/cm^2.

$$J = \frac{I}{A}$$
(3.16)

Current density, J, is defined as the amount of charge that passes through an area per unit time [3]. We can calculate the current density by starting with the concentration of holes and electrons, multiplying by the charge of an electron to convert the units to Coulombs, and multiplying by velocity to get the number of electrons and holes that pass through the cross-sectional area per time: Written out:

$$J_{\text{drift}} = q v_{d,p} p_0 - q v_{d,n} n_0$$
(3.17)

where $v_{d,n}$ is the **electron drift velocity** and $v_{d,p}$ is the **hole drift velocity**. This equation assumes that q is always a positive number. The equation uses a negative sign in front of the electron term to account for the fact that electrons are negatively charged. The **drift velocity** is the average velocity of the electrons or holes due to an electric field.

Later, in the chapter on diodes, we will find out that there is another form of current flow in semiconductors called **diffusion current density**. We label the **drift current density** with the subscript 'drift' at this time to avoid confusion in the future.

Notice that the current density in a semiconductor is due to both electrons and holes. This is different than a conductor, where the current

---

[3] Don't let the symbol 'A' confuse you. If the symbol occurs in an equation, then A refers to area. If the symbol occurs in the units, then A is an amp.

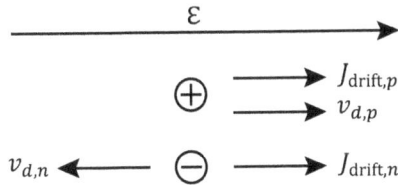

Figure 3.8: Holes move in the same direction as the electric field. Electrons move in the opposite direction as the electric field. The current density is always in the same direction as the electric field.

density is due only to electrons. In fact, in a pure semiconductor, the electrons and holes may contribute equally to current density.

Consider Figure 3.8, where an electric field is pointed to the right ($\vec{\mathcal{E}}$ is positive). The electrons have negative charge, so there is an electrostatic force on the electrons pointed to the left. The holes have a positive charge, so there is an electrostatic force on the holes pointed to the right. Thus, the electrons will move to the left (negative velocity) and the holes will move to the right (positive velocity). Looking at Equation 3.17, we see that the holes contribute to a positive current density (first term), and that the electrons also contribute to a positive current density (second term) because the negative velocity cancels the negative sign in the equation. Here is a very important fact:

> The **drift current density** is always in the direction of the electric field. It does not matter whether the carrier is an electron or a hole.

3.9 Carrier Mobility

The drift velocity of electrons and holes as a function of electric field have been measured for various semiconductors. It has been found to be a function of the temperature, doping concentration, and number of defects in the semiconductor. The drift velocity is usually found to look similar to Figure 3.9. If you look at an individual electron, it may be moving in any direction with any speed. The drift velocity is the average speed of all the electrons or holes due to an electric field. The absolute value signs are used in Figure 3.9 so that the electron drift

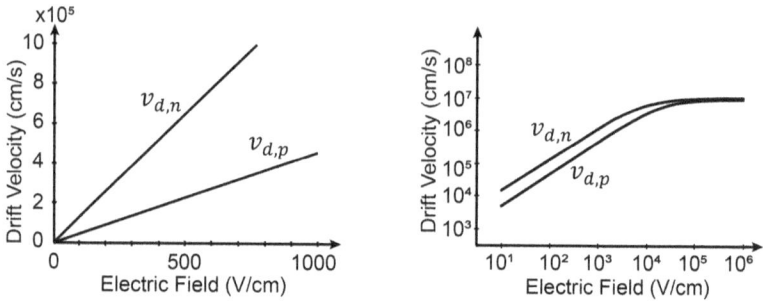

Figure 3.9: Drift velocity of electrons and holes in silicon. (a) At small electric fields, the drift velocity is linear with respect to electric field. (b) The log scale plot shows that the drift velocity saturates at high electric field.

velocity may be plotted in the same quadrant as the hole drift velocity; electrons should have a negative velocity for a positive electric field.

As can be seen in Figure 3.9, the electrons and holes do not accelerate forever, attaining infinite velocities. The electrons and holes interact with all the other charges present, including the other electrons in the conduction band, electrons in the valence band, and even the positive nuclei of every atom.

> An analogy is trying to drive a bumper car at an amusement park to an exit. You start driving, but then your direction and speed randomly changes with every collision. You eventually reach your destination, but your average travel speed is much slower than your maximum instantaneous speed.
>
> A second analogy would be to consider a raindrop falling from a cloud. The raindrop is the electron or hole, and gravity is the electric field. As the raindrop falls due to the force of gravity, it feels the drag of the air molecules around it, and the velocity reaches a maximum velocity, called the terminal velocity. If the raindrop were on a different planet, with a different gravitational pull, the terminal velocity would be different.

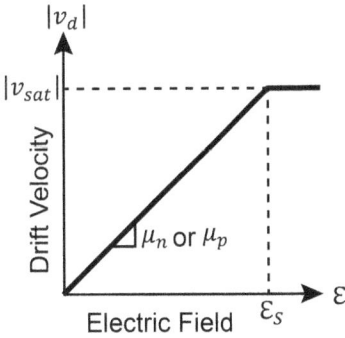

Figure 3.10: Simplified version of the drift velocity versus electric field showing the electron and hole velocity saturation.

From Ohm's law, we expect that as the electric field increases, the current density should increase. However, the current density only increases for low to moderate electric fields. Figure 3.10 shows a simplified relationship between the electron or hole drift velocity and the electric field. The drift velocity is linearly related to the electric field below a specific point, called the **saturation electric field**. Above the saturation electric field, the drift velocity is saturated. Mathematically, this is written as:

$$|v_{d,n,p}| = \begin{cases} \mu_{n,p}\mathcal{E} & \mathcal{E} < \mathcal{E}_S \\ v_{sat,n,p} & \mathcal{E} > \mathcal{E}_S \end{cases} \tag{3.18}$$

where $\mathcal{E}$ is the electric field, $\mathcal{E}_S$ is the saturation electric field, v_{sat} is the **saturation drift velocity**, and μ is the mobility. From Figure 3.9, the saturation electric field is typically in the range of 10^4 to 10^5 V/cm and the saturation drift velocity is typically on the order of 10^7 cm/s. From the graph, we see that the **mobility** is defined as:

$$\mu \equiv \left| \frac{dv_d}{d\mathcal{E}} \right| \tag{3.19}$$

Values for μ_n, μ_p, $\mathcal{E}_{S,n}$, $\mathcal{E}_{S,p}$, $v_{sat,n}$, and $v_{sat,p}$ may be found in Appendix B. The mobilities must be calculated based on the doping concentration in the semiconductor. All of these values depend on the type of semiconductor.

The mobility is a very useful parameter and has units of $\left(\frac{cm^2}{V\cdot s}\right)$. Most of the equations in this book will be written in terms of the mobility, which leads to an important fact. **An equation with mobility in it is only valid if the electric field is below the saturation electric field**

($\varepsilon < \varepsilon_s$). If the electric field exceeds the saturation electric field, the concept of mobility is not valid and you may not use an equation with mobility in it. If in doubt, you must calculate the electric field to determine if you can use the equation.

Example 3.4: Electric fields in excess of 10^5 V/cm are common in semiconductors. Consider a transistor with a minimum feature size of 20 nm and a 1 V supply voltage. (Even smaller features are commonly manufactured.) What is the electric field?

Answer:
$$\varepsilon = \frac{\Delta V}{\Delta x} = \frac{1 \text{ V}}{20 \text{ nm}} = \frac{1 \text{ V}}{20 \times 10^{-7} \text{ cm}} = 5 \times 10^5 \text{ V/cm}$$

The mobility depends upon all the defects in the semiconductor. Measuring the mobility is one way of determining how perfect a semiconductor is. The higher the semiconductor, the fewer the defects. All dopant atoms are defects. The mobility is reduced when dopant atoms are introduced. Appendix B provides equations and shows graphs that provide the mobility as a function of dopant concentration for silicon and other common semiconductors. Appendix B also contains graphs carrier drift velocity versus electric field information. These are all based on experimental values, and thus there is some disagreement among researchers, especially for newer semiconductors.

Sometimes a single value is reported for the mobility of a semiconductor that represents the low-field mobility. The mobility has a maximum value for a pure semiconductor and drops when impurities or defects are present. When a single mobility value is reported, it is reported so that an easy comparison between different semiconductor materials may be made. This single value cannot, in general, be used for designing a semiconductor device. For determining the performance of a semiconductor device, and for every homework assignment, it will ALWAYS be necessary to find a mobility that relates to the doping concentration unless the mobility is given in the problem statement.

Example 3.5: Calculate the electron mobility and hole mobility for silicon doped with $N_A = 10^{16} \text{cm}^{-3}$.

Solution: The equation for mobility from Appendix B is:

$$\mu = \mu_{min} + \frac{\mu_{max} - \mu_{min}}{1 + \left(\dfrac{N}{N_{ref}}\right)^{\gamma}}$$

Looking up parameters from Appendix B for the electron mobility:

$$\mu_{min} = 88.3 \ \frac{\text{cm}^2}{\text{V·s}}$$

$$\mu_{max} = 1330.3 \ \frac{\text{cm}^2}{\text{V·s}}$$

$$N_{ref} = 1.295 \times 10^{17} \text{cm}^{-3}$$

$$\gamma = 0.891$$

$$N = 10^{16} \text{cm}^{-3}$$

For the hole mobility:

$$\mu_{min} = 54.3 \ \frac{\text{cm}^2}{\text{V·s}}$$

$$\mu_{max} = 461.2 \ \frac{\text{cm}^2}{\text{V·s}}$$

$$N_{ref} = 2.35 \times 10^{17} \text{cm}^{-3}$$

$$\gamma = 0.88$$

$$N = 10^{16} \text{cm}^{-3}$$

The electron and hole mobility are then:

$$\mu_n = 88.3 + \frac{1330.3 - 88.3}{1 + \left(\dfrac{10^{16}}{1.295 \times 10^{17}}\right)^{0.891}} = 1215 \ \frac{\text{cm}^2}{\text{V·s}}$$

$$\mu_p = 54.3 + \frac{461.2 - 54.3}{1 + \left(\dfrac{10^{16}}{2.35 \times 10^{17}}\right)^{0.88}} = 437 \; \frac{cm^2}{V \cdot s}$$

Of course, we could have just looked in Table B.3 to obtain these values, but this example shows how the calculation is performed for doping concentrations that are not on the table. Notice that the mobility values are just a little lower than the mobility for a pure semiconductor given in Table B.1. Also, note that we used the same doping concentration, the acceptor doping concentration, to calculate both the electron mobility and hole mobility. This is because the mobility is determined by any impurities in the semiconductor, including dopant atoms, whether they be acceptor or donor atoms.

Example 3.6: Calculate the saturation electric field for electrons and holes in silicon with an acceptor doping concentration of 10^{16} cm^{-3}.

Solution. The velocity saturation for electrons and holes are given in Appendix B:

$$v_{sat,n} = 1.05 \times 10^7 \; cm/s$$

$$v_{sat,p} = 0.94 \times 10^7 \; cm/s$$

The electron mobility and hole mobility were found in Example 3.5 to be $\mu_n = 1215 \; \frac{cm^2}{V \cdot s}$ and $\mu_p = 437 \; \frac{cm^2}{V \cdot s}$. Using Figure 3.10, the saturation electric field is:

$$\mathcal{E}_{S,n} = \frac{v_{sat,n}}{\mu_n} = \frac{1.05 \times 10^7 \; cm/s}{1215 \; \frac{cm^2}{V \cdot s}} = 8.6 \; kV/cm$$

$$\mathcal{E}_{S,p} = \frac{v_{sat,p}}{\mu_p} = \frac{0.94 \times 10^7 \; cm/s}{437 \; \frac{cm^2}{V \cdot s}} = 21.5 \; kV/cm$$

3.10 Current Density, Resistivity, Conductivity, and Ohm's Law

In most cases, we will assume that the electric fields are sufficiently small that the electrons and holes have not reached their saturation velocities. In this section, we will find new equations for the current density based on this assumption. Using the equation for mobility, solving for drift velocity, and inserting that into the current density equation, we get:

$$J_{drift} = q\mu_p p_0 \mathcal{E} + q\mu_n n_0 \mathcal{E} \tag{3.20}$$

where we had to change the sign of the second term because the electrons move in opposite direction of the electric field. If we factor out the electric field, we can find a term that represents the **conductivity** of the semiconductor:

$$\begin{aligned} J_{drift} &= \left(q\mu_p p_0 + q\mu_n n_0\right)\mathcal{E} \\ &= \sigma\mathcal{E} \end{aligned} \tag{3.21}$$

For conductors, the conductivity is a material property. For semiconductors, the conductivity can be changed by doping the semiconductor which in turns changes the electron and hole concentrations. The conductivity has units $\left(\frac{1}{\Omega\cdot cm}\right)$, and is equal to:

$$\sigma = q\mu_p p_0 + q\mu_n n_0 \tag{3.22}$$

Resistivity is $1/\sigma$, and has units of $(\Omega \cdot cm)$. The resistivity is equal to:

$$\varrho = \frac{1}{\sigma} = \frac{1}{q\mu_p p_0 + q\mu_n n_0} \tag{3.23}$$

Examples 3.1 and 3.2 show that the majority carrier concentration is orders of magnitude larger than the minority carrier concentration. Thus, when finding the conductivity and resistivity, only the term with the majority carrier needs to be calculated, and the term with the minority carrier concentration can be neglected. For example, in n-type material (doped with donors), $n_0 \gg p_0$ and we obtain:

n-type semiconductor:
$$\sigma \approx q\mu_n n_0 \tag{3.24}$$

$$\varrho \approx \frac{1}{q\mu_n n_0} \tag{3.25}$$

The opposite occurs for a p-type semiconductor:

p-type semiconductor:

$$\sigma \approx q\mu_p p_0 \tag{3.26}$$

$$\varrho \approx \frac{1}{q\mu_p p_0} \tag{3.27}$$

Typically, in a circuits course we learn Ohm's law to be:

$$V = IR \tag{3.28}$$

When working with semiconductors, we can work with an equivalent form of Ohm's Law in terms of the material properties:

$$\mathcal{E} = J_{\text{drift}}\, \varrho \tag{3.29}$$

where:

$$V = \mathcal{E}L \tag{3.30}$$

$$I = JA \tag{3.31}$$

$$R = \frac{\varrho L}{A} \tag{3.32}$$

where A is the cross-section area through which the current flows and L is the length of the semiconductor.

In this section we saw that we could write simple equations for current density, resistivity, and conductivity based on the concept of mobility. If the electric field exceeds $\mathcal{E}_S$, the electric field at which the carrier velocity saturates, then we cannot properly define a resistivity or conductivity. To calculate the current density in this case, we have to use the form of the equation with the electron or hole drift velocity explicitly defined.

> **The concepts of resistivity and conductivity only make sense when the carriers are not velocity saturated. That is, only for $\mathcal{E} < \mathcal{E}_S$.**

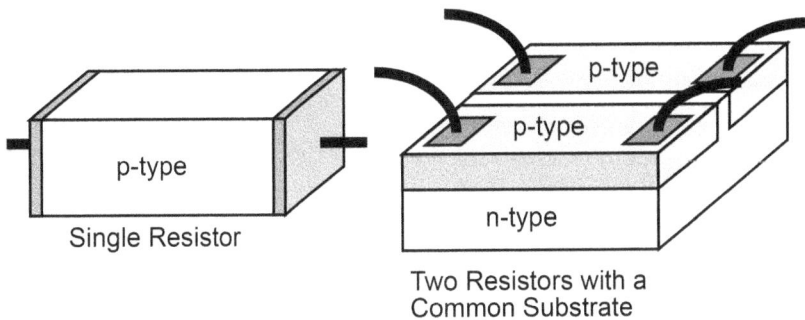

Figure 3.11: (left) A single resistor made using a p-type semiconductor. The ends of the resistor are made of metal to permit wires to be attached. (right) Two resistors are made on a single semiconductor substrate. The resistors are p-type, and the substrate is n-type, providing electrical isolation between the resistors. In this case, the metal and wires are formed on top of the substrate.

3.11 Diffused Resistors

Thus far, we have discussed uniformly doped semiconductors. In practice, it is desired to put more than one device on a single semiconductor chip, or die. For example, there are more than 1 billion transistors on a modern memory chip or microprocessor chip. How can we do this? Consider first a single resistor made out of a semiconductor, as shown in Figure 3.11a. This is the kind of resistor we considered in the last section. Now, let us be clever. Let us assume, for example, that we want two resistors that are doped p-type on the same chip. To do this, we will start with a uniformly doped n-type wafer and **diffuse** a p-type dopant onto the top of the wafer in two regions. That is, we will make a region of the wafer p-type using a process in which dopant atoms diffuse from the top surface of the semiconductor into the body. A 3D image of the two resistors in a single die is shown in Figure 3.11b. Now we have two isolated resistors. Current will not travel from a p- region to an n- region unless a voltage is applied between the two regions. The details of the physics of a p- region adjacent to an n- region will be delayed until Chapter 6.

I want to explain the term **diffuse** a little better because it is commonly used by engineers who fabricate semiconductor devices. For our example, we will put boron, a p-type dopant atom, on the surface of the semiconductor substrate. The wafer is then heated to a very high temperature, around 1000°C, for some minutes or hours. The boron atoms will slowly diffuse through the silicon, moving from the surface to a point further into the silicon. This process is called diffusion. This process is similar to dropping a single drop of dye into a glass of water. In a bit, the dye has spread further into water, and eventually the entire glass of water has the dye in it. In our case, we stop the diffusion process by cooling the wafer to room temperature before the dopant atoms spread uniformly throughout the silicon wafer. By controlling the temperature and time, we can control how deep the boron diffuses into the semiconductor. There are many methods of introducing dopant atoms to the substrate and the details may be found in a book on microfabrication techniques.

With diffused resistors, the thickness of the resistor is the same everywhere because the temperature and time of the diffusion is the same for all resistors. This procedure is commonly called batch fabrication because we can make 10 or 10,000,000 resistors on a semiconductor wafer simultaneously. We can put the initial boron where we want on the wafer, and hence we have control over the individual location, length, and width of all the diffused resistors; just not individual control over the depth.

Figure 3.12 shows a plot of the doping concentration as a function of depth. In Figure 3.12a is an ideal scenario, while Figure 3.12b shows that the doping concentration typically varies with distance with the largest dopant concentration located at the semiconductor surface.

3.12 Sensitivity Analysis

It is common to perform a sensitivity analysis to understand how sensitive a parameter is to the underlying parameters. Let us call the parameter of interest the output, and the parameters that are varied are the inputs. For example, if we are designing a resistor, the output may be the resistance and the inputs may be length, width, thickness, resistivity, or doping concentration. We find the sensitivity of the

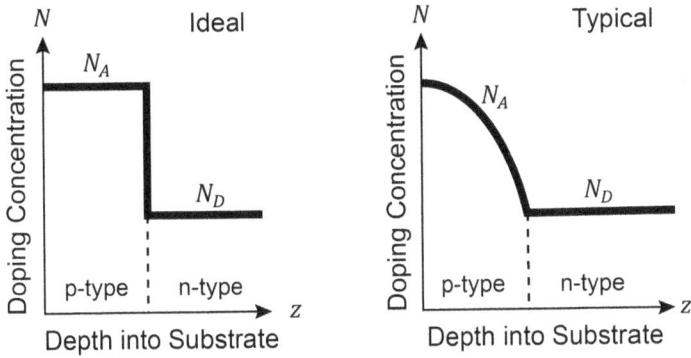

Figure 3.12: These plots show the doping concentration as a function of distance into the substrate from the top surface. (left) The doping concentration of the diffused region (N_A) is ideally flat. (right) It is common for the diffused doping concentration to have a non-linear doping profile.

output with respect to a single input at a time. Using a percent change, the sensitivity is defined as:

$$S_{x_1,x_2} = \frac{1}{x_1}\frac{\partial x_1}{\partial x_2} \times 100\% \qquad (3.33)$$

where x_1 is the output and x_2 is an input. The units for sensitivity depend upon the parameters of interest.

We will use the sensitivity to understand how the input affects the output. To make this relation clear, let us assume a linear relationship between the output and input and change the derivative in Equation 3.33 to a ratio of differences. After rearranging terms, we get:

$$\frac{\Delta x_1}{x_1}(\%) = S_{x_1,x_2}\Delta x_2 \qquad (3.34)$$

This equation allows us to understand how an input affects the output using the sensitivity. If there are three inputs, we could find out which input has the greatest effect on the output.

Example 3.7: The resistance of a resistor is given in Equation 3.32. (a) Derive an equation for the sensitivity of the resistance with respect to length. (b) Consider a silicon resistor with a resistivity of 1 Ω·cm, length of 1 mm, width of 10 μm, and depth of 2 μm. What is the sensitivity of resistance with respect to length? (c) If the length varies by ± 1 μm, what is the percent change in resistance?

Answer: (a) Using Equation 3.32 and 3.33, the sensitivity of resistance with respect to length is:

$$S_{R,L} = \frac{1}{R}\frac{dR}{dL} = \frac{A}{\varrho L}\frac{d\left(\frac{\varrho L}{A}\right)}{dL} \times 100\% = \frac{100\%}{L}$$

Notice that the resistivity, width, and thickness do not matter for this sensitivity calculation.

(b) Inserting the dimensions, which in this case is only the length, the sensitivity of resistance with respect to length becomes:

$$S_{R,L} = \frac{100\%}{1\ \text{mm}} = 100\ \text{mm}^{-1}\ \%$$

(c) The percent change in resistance is found from Equation 3.34:

$$\frac{\Delta R}{R}(\%) = S_{R,L}\Delta L = 100\ \text{mm}^{-1}\% \times 1\mu\text{m} \times \frac{1\ \text{mm}}{10^3\ \mu\text{m}} = 0.1\ \%$$

3.13 Temperature Effects

The effect of temperature on a semiconductor is very important because we want our semiconductor devices to work at all temperatures. For example, we want our cell phone to work indoors and outdoors, whether it be summer or winter. We also want the microprocessor that runs a car engine to work on a hot summer day, despite being located under the hood of the car next to a very hot engine.

3.13.1 Effect of temperature on carrier concentration

The intrinsic carrier concentration, n_i, is very strongly dependent on temperature. The equation for n_i is given in Equation 3.10, where the term $\exp\left(-\frac{E_G}{2k_BT}\right)$ explicitly shows the temperature dependence.

However, both N_C and N_V also depend on temperature. If the effective density of states are known at 300 K, as is commonly reported and as shown in Appendix A, then they may be found at any temperature using the following equations:

$$N_C(T) = N_C \left(\frac{T}{300}\right)^{3/2} \qquad (3.35)$$

$$N_V(T) = N_V \left(\frac{T}{300}\right)^{3/2} \qquad (3.36)$$

where the temperature used must be in Kelvin. Now it is possible to calculate the intrinsic carrier concentration as a function of temperature using Equation 3.10:

$$n_i = \sqrt{N_C(T)N_V(T)\exp\left(-\frac{E_G}{k_B T}\right)} \qquad (3.37)$$

It ought to be noted that the band gap, E_G, also changes with temperature, but the band gap variation is small and is a second order correction. The intrinsic carrier concentration as a function of temperature of some semiconductors are shown in Figure 3.13. Notice that the larger the band gap, the smaller n_i is. Also notice that the intrinsic carrier concentration increases rapidly with temperature.

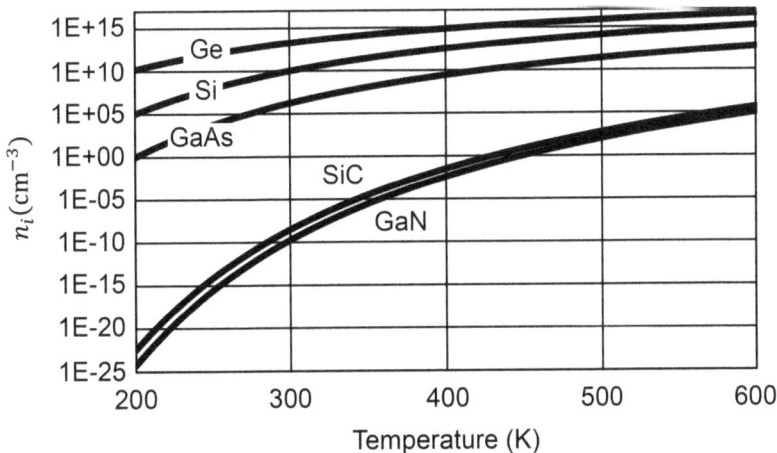

Figure 3.13: Intrinsic carrier concentration versus temperature for five different semiconductors.

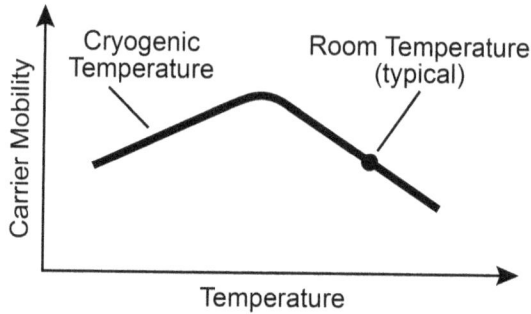

Figure 3.14: Relationship between mobility and temperature in a semiconductor. At low temperatures, the mobility increases with increasing temperature. At high temperatures, the mobility decreases with increasing temperature. Room temperature is typically on the right side where the mobility decreases with increasing temperature.

The majority carrier concentration does not change with temperature because the majority carrier concentration is set by the doping concentration. The minority carrier concentration increases as the temperature increases because n_i increases with temperature.

3.13.2 Effect of temperature on mobility

The electron and hole mobility depend upon temperature. The values listed in Appendix C are all for room temperature. The mobility as a function of temperature generally looks like Figure 3.14, but around room temperature we are *typically* on the right side of the curve where the mobility is decreasing with increasing temperature. The word *typical* is important here. The shape of the mobility curve with respect to temperature depends upon the type of semiconductor, dopant concentration, type of dopant used, and often other details of the defects in the semiconductor. Thus, in some cases, we could be on the left side of the mobility curve at room temperature, where the mobility is increasing with temperature.

3.13.3 Temperature Coefficient of Resistance

The **temperature coefficient of resistance**, abbreviated **TCR**, describes how a change in temperature changes the resistance of a resistor with the assumption that the change is linear in the temperature range of interest. Ignoring higher order terms, the following equation describes how the resistance changes as a function of temperature:

$$R = R_0 + R_0 \alpha \Delta T + H.O.T. \cong R_0 (1 + \alpha \Delta T) \qquad (3.38)$$

where R_0 is the resistance at a specified reference temperature and α is the **temperature coefficient of resistance**. This equation may be written as:

$$\frac{R - R_0}{R_0} = \frac{\Delta R}{R_0} = \alpha \Delta T \qquad (3.39)$$

The TCR has units of 1/temperature (K^{-1}). If the resistance increases with temperature, then TCR is positive. If the resistance decreases with temperature, then TCR is negative.

It is commonly said that "metals have a positive TCR; semiconductors have a negative TCR." I have seen this statement in many places. Additionally, I have seen lab experiments designed where the student measures the TCR of a sample to determine whether it is a metal or semiconductor. Sounds simple, right? In reality, **a semiconductor may have a positive or negative TCR**. Thus, the common statement is false, and the lab experiment is not a good experiment.

There are two primary factors that affect the conductivity (or resistivity) of a semiconductor: the mobility and the carrier concentration. An increase in the temperature causes the carrier concentration to increase. This in turn causes the resistance to decrease with an increasing temperature. An increase in temperature may increase or decrease the mobility, and thus the resistance may increase or decrease due to the change in mobility. In the end, the combined effect of changing carrier concentration and changing mobility could cause the resistance to increase or decrease with increasing temperature. Thus, we cannot make a general statement about the sign of the TCR for a semiconductor.

Some general trends may be observed. In a lightly doped semiconductor, it is easier to change the electron and hole concentration and therefore it is more likely to have a negative TCR. In a heavily doped semiconductor, the majority carrier is determined by the dopant concentration and the minority carrier concentration is small. Therefore, the change in mobility is more likely to dominate and, if we are on the right side of Figure 3.14, the TCR is more likely to be positive. Keep in mind that there are many examples that break these generalizations.

Example 3.8: Silicon has a doping concentration of $N_A = 10^{15}$ cm^{-3}:

What is the:	Solution:
Hole concentration:	$$p_0 = N_A = 10^{15} \text{ cm}^{-3}$$
Electron concentration:	$$n_0 = \frac{n_i^2}{p_0} = \frac{10^{20} \text{ cm}^{-6}}{10^{15} \text{ cm}^{-3}} = 10^5 \text{ cm}^{-3}$$ Using n_i from Appendix A.
Electron mobility	$$\mu_n = 1330 \frac{\text{cm}^2}{\text{V·s}}$$ Using Appendix B.
Hole mobility	$$\mu_p = 460 \frac{\text{cm}^2}{\text{V·s}}$$ Using Appendix B.
Semiconductor type (n or p)	**p-type** because there are more holes than electrons.
Conductivity	$$\sigma = q\mu_p p_0$$ $$= (1.6 \times 10^{-19} \text{ C}) \left(460 \frac{\text{cm}^2}{\text{V·s}}\right) (10^{15} \text{ cm}^{-3})$$ $$= 0.0736 \left(\frac{1}{\Omega \cdot \text{cm}}\right)$$ Be careful when evaluating the units.
Resistivity	$$\varrho = \frac{1}{\sigma} = \frac{1}{0.0736} \, \Omega \cdot \text{cm} = 13.6 \, \Omega \cdot \text{cm}$$
Approx. where is the Fermi level?	Near the valence band because this semiconductor is doped p-type.
Numerically, where is the Fermi level?	$$E_F - E_V = -kT \ln \frac{p_0}{N_V}$$ $$= -0.0259 \text{ eV} \ln \frac{10^{15} \text{ cm}^{-3}}{1.87 \times 10^{19} \text{ cm}^{-3}}$$ $$= 0.255 \text{ eV}$$

Example 3.9: It is desired to obtain a resistivity of $0.1 \ \Omega \cdot \text{cm}$ in GaN using an n-type dopant. What should the doping concentration be?

Answer: The first step is to identify the equations we will need. Let's start with the equation for resistivity, and use the approximation that $n_0 \gg p_0$ in order to simplify the equation:

$$\varrho = \frac{1}{\sigma} = \frac{1}{q\mu_n n_0} = 0.1 \ \Omega \cdot \text{cm}$$

Since q is a constant, there are two unknowns. An equation for the electron mobility may be found from Appendix B in terms of n_0 by using the approximation $N_D = n_0$:

$$\mu_n = 100 + \frac{1600 - 100}{1 + \left(\frac{N_D}{3 \times 10^{17}}\right)^{0.7}} \ \frac{\text{cm}^2}{\text{V} \cdot \text{s}}$$

Using $N_D = n_0$, these two equations may be combined:

$$\frac{1}{\varrho} = \frac{1}{0.1 \ \Omega \cdot \text{cm}} = q \left[100 + \frac{1600 - 100}{1 + \left(\frac{N_D}{3 \times 10^{17}}\right)^{0.7}} \right] N_D$$

There is only one unknown, but this is difficult to solve by hand. Use a computer to numerically solve this equation. The answer is:

$$N_D = 4.9 \times 10^{16} \ \text{cm}^{-3}$$

3.14 Summary of Key Points

- A semiconductor has a conduction band and a valence band separated in energy by a value called the band gap.
- The band gap is a material property. It cannot be changed with voltage or doping.
- There are no quantum states with an energy within the band gap that an electron may occupy.
- The top of the valence band is labeled E_V.
- The bottom of the conduction band is labeled E_C.
- Electrons occupy the conduction band.

- Electrons mostly fill the valence band, but we will refer to the holes in the valence band instead of the electrons.
- The effective density of states tells us how many quantum states exist for electrons in the conduction band and holes in the valence band. These quantum states may or may not be filled with electrons and holes.
- The Fermi-Dirac and Maxwell-Boltzmann distribution functions tell us the probability of a quantum state being filled with an electron as a function of energy, relative to the Fermi level.
- The electron concentration and hole concentration of an intrinsic semiconductor are equal to each other.
- The intrinsic carrier concentration, n_i, may be considered a material property. Note that it is very temperature dependent.
- $n_0 p_0 = n_i^2$ is a very useful equation to memorize. It will help you solve many semiconductor problems throughout the book.
- Semiconductors may be doped with acceptor atoms or donor atoms.
- Dopant atoms do not change the net charge within the semiconductor. The overall charge will be zero, although we will see in future chapters how to change the *local* distribution of the charges.
- Acceptor atoms increase the hole concentration in the valence band.
- Donor atoms increase the electron concentration in the conduction band.
- The middle of the band gap is labeled E_i.
- The drift current density is determined by the electric field. The equation for the drift current density using mobility is analogous to Ohm's law for circuit analysis.
- Mobility is the slope when plotting the carrier velocity versus electric field.
- When the carrier drift velocity equals the saturation velocity, the concept of mobility may not be used.
- Electrons in the conduction band have negative charge.
- Holes in the valence band have positive charge.
- The drift current always flows in the direction of the electric field. This is true for both electrons and holes.

3.15 Problems

There are many problems here, but additional permutations may be easily created. In any problem below, easy permutations made by made by (1) substituting one type of semiconductor for a new type of semiconductor, (2) changing the doping concentration to a new value within the range 10^{14}cm^{-3} to 10^{19}cm^{-3}, or (3) changing from n-type to p-type. For example, in problem #1, change silicon to any of Ge, GaN, GaAs, or SiC.

1. Name three acceptor dopant elements for silicon.

2. Name three donor dopant elements for silicon.

3. (trick question) What kind of dopant is silicon in GaAs?

4. Draw a band diagram of a semiconductor, showing the valence band, the conduction band, and a band gap of 1.5 eV. The y-axis should be energy.

5. A semiconductor has a band gap of 1.2 eV. What is the energy in Joules?

6. Silicon is doped with acceptor atoms at a concentration of $N_A = 10^{18} \text{cm}^{-3}$. What are (a) the electron concentration and (b) the hole concentration.

7. GaAs is doped with acceptor atoms at a concentration of $N_A = 10^{18} \text{cm}^{-3}$. What are (a) the electron concentration and (b) the hole concentration.

8. Sketch the Fermi-Dirac distribution function as a function of energy. Indicate the Fermi energy, E_F, and the value of the Fermi-Dirac distribution at the Fermi energy.

9. Draw a band diagram for an n-type semiconductor, including the Fermi level.

10. Draw a band diagram for a p-type semiconductor, including the Fermi level.

11. GaN is doped with donor atoms at a concentration of $N_D = 10^{16} \text{cm}^{-3}$. What are:
    a. the electron concentration
    b. the hole concentration

    c. $E_F - E_i$
    d. $E_F - E_V$
    e. $E_C - E_F$
    f. Draw the band diagram and indicate the conduction band, valence band, intrinsic level, and Fermi level. Label the numerical values computed above, and also indicate the band gap.

12. Using a computer and your favorite mathematical program, plot the Fermi-Dirac distribution function as a function of energy. On the same plot, show the Maxwell-Boltzmann distribution function. The y-axis should range from 0 to 2. The x-axis should range from $E_F - 1$ eV to $E_F + 1$ eV. Do this calculation at a temperature of 300 K. Over what energy range does the Maxwell-Boltzmann distribution function approximate the Fermi-Dirac distribution function.

13. Using a computer and your favorite mathematical program, plot the Fermi-Dirac distribution function as a function of energy. The y-axis should range from 0 to 1. The x-axis should range from $E_F - 0.2$ eV to $E_F + 0.2$ eV. Do this calculation for three temperatures: 77 K, 300 K, 400 K.
    a. Create a plot with all three curves superimposed.
    b. If we consider an energy of $E_F + 0.1$ eV with a number of quantum states available at that energy, does the electron concentration at that energy increase or decrease as the temperature increases?
    c. If we consider an energy of $E_F - 0.1$ eV with a number of quantum states available at that energy, does the hole concentration at that energy increase or decrease as the temperature increases?

14. Use a computer and calculate the resistivity of n-type silicon for a doping concentration ranging from $N_D = 10^{14}$ cm^{-3} to a doping concentration of $N_D = 10^{19}$ cm^{-3}. Show your work to derive the resistivity, and show what numbers you use in the equations. Using the computer, make a nice plot showing the resistivity. Use a log-log axes. Label your axes, including units. Keep in mind the mobility is a function of doping concentration.

Note: You can easily find graphs of resistivity versus doping concentration to check your answer.

15. Use a computer and calculate the current density of n-type silicon for a doping concentration ranging from $N_D = 10^{14}$ cm^{-3} to a doping concentration of $N_D = 10^{19}$ cm^{-3}. Show your work to derive the current density in (A/cm^2), and show what numbers you use in the equations. Using the computer, make a nice plot showing the current density. Use a log-log axes. Label your axes, including units. The silicon is 100 µm long, and 1 V is applied across the silicon.

16. Use a computer and calculate the current density of n-type silicon for a doping concentration of $N_D = 10^{16}$ cm^{-3}. The length of the silicon varies from 10 nm to 1 mm. Show your work to derive the current density, and show what numbers you use in the equations. Using the computer, make a nice plot showing the current density. Use a log-log axes. Label your axes, including units. 1 V is applied across the silicon. Watch out for velocity saturation effects!

17. Use a computer and plot the intrinsic carrier concentration as a function of temperature for GaAs. Use a log scale on the y- axis, and use a temperature range from 200 K to 600 K. Label your axes, including units. You can check your answer by comparing to Figure 3.13.

18. The channel region of a MOSFET in a modern IC may be only 50 nm long.
    a. Assuming the transistor is 100 nm wide, and the depth is 5 nm, how many silicon atoms make up the channel?
    b. For a doping concentration of 10^{17} cm^{-3} Boron atoms, how many Boron atoms are in this volume?
    c. How many holes are in this volume?
    d. How many electrons are in this volume?

    Note: These questions require you to find a concentration and multiply by the volume to arrive at the actual number of particles.

19. SiC is doped with $N_A = 10^{18}$ cm^{-3}. What are:
    a. The electron concentration

   b. The hole concentration

   c. Without performing a calculation, is E_F near E_C or is E_F near E_V?

   d. What is $E_F - E_V$?

   e. What is the conductivity?

   f. What is the resistivity?

   g. What is the electron drift velocity for an electric field of 100 V/cm?

   h. What is the hole drift velocity for an electric field of 100 V/cm?

20. A GaAs resistor is doped with donor atoms at a concentration of $N_D = 10^{15} \text{cm}^{-3}$. The resistor is 1 mm long, 100 μm wide, and 2 μm thick. What are:

   a. The resistivity

   b. The resistance

   c. The current when 10 V is applied.

21. Consider silicon with a doping concentration of $N_A = 10^{15} \text{ cm}^{-3}$. What electric field causes the electrons to reach their saturation velocity?

22. A silicon resistor is doped with donor atoms at a concentration of $N_D = 10^{15} \text{ cm}^{-3}$. The resistor is 10 μm long, 2 μm wide, and 1 μm thick. What are:

   a. The resistivity

   b. The resistance

   c. The current when 1 V is applied.

   d. The current when 100 V is applied. Be careful on this question!

23. It is desired to have a resistor with a value of 100 Ω. Design a resistor using GaN. Determine values for the doping concentration, length, width, and thickness. The constraints are the doping concentration must be within the range 10^{14} cm^{-3} to 10^{19} cm^{-3} and the minimum geometry dimension is 10 μm.

24. Consider a diffused n-type resistor into a p-type silicon substrate.

   a. Draw a top view of the device. Include the metal contacts in your drawing.

b. Draw a cross-section of the device, cut through the length of the resistor. Include the metal contacts on your drawing.

25. Figure 3.12b shows that the doping concentration varies with distance. When the diffusion is performed using what is known as a **"limited-source diffusion"**, the doping concentration is represented by the following equation:

$$N_A(y) = \frac{Q}{\sqrt{\pi Dt}} \exp\left(-\frac{y^2}{4Dt}\right)$$

where Q is the doping concentration at the surface and the product Dt is a parameter that controls how deep the profile extends.

a. Find the value of y where $N_A(y) = N_B$, where N_B is the background doping concentration of the wafer. Let this value be H to represent the height of the resistor.

b. If the value of N_B is increased, does the height of the resistor increase or decrease?

c. Find the conductivity, $\sigma(y)$, as a function of y. Assume that the mobility is constant and independent of doping concentration throughout the thickness. Replace p with expressions based on the doping concentration, N_A, and ignore the term with n because that is a minority carrier concentration whose contribution is small.

d. Find the conductance of the resistor from:

$$G = \frac{W}{L} \int_0^H \sigma \, dy$$

As you solve this problem, replace the integral with the definition of the error function, erf (). Your answer should be in terms of erf ().

e. Find the resistance of the resistor from $R = 1/G$.

26. **Sensitivity Analysis.** In this chapter, the sensitivity of a resistor with respect to its length was calculated. Let us examine the sensitivity of an n-type resistor in more detail.

a. What is the resistance sensitivity with respect to width, where the cross-sectional area is $A = WH$?

b. What is the resistance sensitivity with respect to doping concentration?

c. Consider a GaAs resistor doped with donor atoms at a concentration of $N_D = 10^{15} \text{cm}^{-3}$. The resistor is 1 mm long, 100 μm wide, and 2 μm thick. What are $S_{R,L}$, $S_{R,W}$, and S_{R,N_D}?

d. Using the results from part (c), if the width can be controlled to ± 1 μm, what is the percent variation in resistance?

e. Now the resistor dimensions are changed to 2 mm long, 200 μm wide, and 2 μm thick. The resistance remains the same. The width is still controlled to ± 1 μm. What is the percent variation in resistance? Is it smaller or larger than before? What is the tradeoff with respect to cost?

4 Energy, Electric Potential, and Voltage

In the last chapter, we worked with the potential energy of an electron, and used the applied voltage to find the current in a semiconductor. Both of these topics really need more attention to clarify these concepts.

There are three concepts to be discussed in this chapter: the **energy** of the electrons and holes, the **electric potential** within the semiconductor, and the **applied voltage**.

4.1 Symbols

The symbols of energy, electric potential, and voltage are:

Energy	E
Electric Potential	ϕ
Applied Voltage	V

There is no such thing as an absolute energy, absolute electric potential, or absolute voltage. These entities are all relative measures. For example, have you ever seen a voltmeter with a single lead? No, because you can only measure the voltage between two nodes on a circuit. The difference between two energies, or two electric potentials, or two applied voltages may be represented as:

$$E_{12} = E_1 - E_2$$

$$\phi_{12} = \phi_1 - \phi_2$$

$$V_{12} = V_1 - V_2$$

Here are a few examples:

$$E_{Fi} = E_F - E_i$$

$$\phi_{SB} = \phi_S - \phi_B$$

$$V_{GS} = V_G - V_S$$

To keep up with convention, it may be necessary to use labels that violate the subscripting scheme described above. For example, E_G represents the band gap energy of a semiconductor, V_T is the threshold voltage of a MOSFET, and V_{FB} is the flat band voltage of a MOSFET.

4.2   Units

Units of **Energy (E)**:

> Joule (J), or
>
> Electron-Volt (eV)

Conversion: $1 \text{ eV} = 1.6 \times 10^{-19} \text{ J}$

The Joule is in SI units, but the electron-volt is not an SI unit. Despite its name, an electron-volt is a measure of energy, not voltage.

Units of **Electric Potential (ϕ)**:

> Volt (V)

The electric potential is a measure of the electron or hole energy divided by charge. That is,

$$\Delta E_C = -q\Delta\phi \qquad \text{(Electrons)}$$

$$\Delta E_C = q\Delta\phi \qquad \text{(Holes)}$$

where we have used E_C as the energy. We could have used E_V, because $\Delta E_C = \Delta E_V$.

Units of **Applied Voltage (V)**:

> Volt (V)

The applied voltage is a measure of the **average** electron energy divided by charge. That is,

$$\Delta E_F = -q\Delta V \tag{4.1}$$

Notice that we defined two variables, ϕ and V, both with units of a Volt, but they mean different things. The electric potential is applicable to the individual electrons and may or may not be measurable. The applied voltage relates to the average electron energy and may be measured.

The applied voltage is an electro-motive force (emf), or "voltage" as it is commonly called, applied using a battery or power supply. The battery or power supply doesn't act on individual electrons. The battery or power supply sets the average potential difference between its terminals. The applied voltage may be measured with a voltmeter. A voltmeter measures the average potential energy difference for electrons between two points in a circuit, divided by charge. The average potential energy difference provides an ability to do **work** on an electron (work in the physics sense), and hence can be measured.

The electric potential is the combination of the applied voltage and the internal electron energy. A voltmeter may only measure the applied voltage.

Because we cannot place a voltmeter within a semiconductor, we will reserve the use of the symbol V for the applied voltage at an electrode, where the voltmeter may be used, and use the symbol ϕ for the electric potential within the semiconductor. Typically, we will use the term **applied voltage** for the voltage on an electrode because we will be interested in how the semiconductor devices operate as a function of the voltage that we apply to the electrodes.

Example 4.1: What is the potential energy of an electron at 1 V relative to ground?

Answer:

$$\Delta E = (-1.6 \times 10^{-19}\,\text{C})(1\,\text{V}) = -1.6 \times 10^{-19}\,J$$

This answer can be converted to units of eV:

$$\Delta E = (-1.6 \times 10^{-19}\,J)\frac{1\,\text{eV}}{1.6 \times 10^{-19}\,J} = -1\,\text{eV}$$

This example illustrates why physicists (and the author) like to use units of electron-volts. When working with electrons and holes, it is

much easier to perform these calculations in one's head when the conversion factor is ± 1, rather than using SI units when one has to work with a conversion factor of $\pm 1.6 \times 10^{-19}$.

Example 4.2: What is the potential energy of an electron at 0.7 V relative to ground?

Answer:

$$\Delta E = -0.7 \text{ eV}$$

Example 4.3: What is the potential energy of a hole at 1 V relative to ground?

Answer:

$$\Delta E = 1 \text{ eV}$$

Example 4.4: What is the electric potential of an isolated electron that has 2 eV potential energy due to an electric field?

Answer:

$$\Delta \phi = -2 \text{ V}$$

The electric potential applies to individual electrons.

4.3   Zero Energy

There is no such thing as an absolute zero energy. Energy is similar to electric potential. There is no absolute zero voltage. A voltage is always defined as the electric potential between two points. One of these points may be a ground, which is an agreed upon reference point. But there are always two leads to a voltmeter because a voltage at one point is defined relative to the voltage at another point. Similarly, energy is always measured as the potential energy of one state relative to another state. A zero energy may be defined, but it is only for our convenience, similar to the use of a ground node in an electric circuit.

Example 4.5: What is the potential energy of a 1 kg brick located 1 m above the ground, relative to the ground, if the zero potential energy is defined at ground level? What is the potential energy of the brick relative to the ground if the zero potential energy is located 10 m above the ground?

Solution: For the case where the zero potential energy is defined at ground level:

$$\Delta E = mg\Delta x = (1 \text{ kg}) \left(9.8 \ \frac{\text{m}}{\text{s}^2}\right)(1 \text{ m} - 0 \text{ m}) = 9.8 \text{ J}$$

For the case where the zero potential energy is defined at 10 m above ground:

$$\Delta E = mg\Delta x = (1 \text{ kg}) \left(9.8 \ \frac{\text{m}}{\text{s}^2}\right)(-9 \text{ m} - -10 \text{ m}) = 9.8 \text{ J}$$

The potential energy is the same.

Example 4.6: What is the kinetic energy of a book sitting on a table?

Answer: The velocity is zero, so the kinetic energy is simply:

$$E_{K.E.} = \frac{1}{2}mv^2 = 0$$

But wait! The table is on the earth, and the earth spins about its axis. Assuming that we are in the mid-continental United States, the table is moving at ~700 mph. Thus, the kinetic energy must be quite large!

But wait again! The earth is moving around the sun! And the sun is orbiting the Milky Way galaxy. And the Milky Way galaxy is part of the Andromeda cluster of galaxies. What is the kinetic energy?

We have to define a zero velocity, known as an inertial reference frame, and do all our calculations from there. This example shows that there is no such thing as an absolute zero kinetic energy.

4.4   Thermal Equilibrium

A system at thermal equilibrium is a system in which the average energy of the atoms, electrons, gas molecules, etc., are at an equilibrium value, and the average energy is uniform. That is, there is no power being delivered to the system, or applied voltage, or applied current, or light, or even temperature distributions within the system. The system has had enough time for any energy exchange to occur, and for the system to come to its final average resting state. The system will have a single temperature value.

In steady state, nothing is changing with time. That is, all time derivatives are equal to zero. A system in thermal equilibrium must be in steady state. A system in steady state does not have to be in thermal equilibrium.

For the analysis of semiconductors, we will typically start with the case of thermal equilibrium. Then an applied voltage will be applied and the system will no longer be in equilibrium. We will assume that the non-equilibrium case is close enough to the thermal equilibrium case that the Fermi-Dirac distribution function and Maxwell-Boltzmann distribution functions are still valid. If we do not make this assumption, then we will not be able to calculate the electron and hole concentrations using the equations from Chapter 3. Non-equilibrium conditions commonly occur, but we will not get into those cases in this book.

Example: Consider a pendulum, such as used in a grandfather clock, swinging (oscillating) at a specific rate. Is it in thermal equilibrium?

Answer: Yes, we can consider the pendulum to be at thermal equilibrium if we ignore any energy losses such as friction. The pendulum changes its energy between potential energy (at its peak height) and kinetic energy (at the bottom of the swing), but the overall energy is does not change.

Example: Consider a semiconductor with electrons in constant motion, running into each other and exchanging energy with other electrons. Is this system in thermal equilibrium?

Answer: Yes. Even though individual electrons may increase or decrease in energy, the overall energy remains the same.

Example: Consider a semiconductor with light shining on it. Is this system in thermal equilibrium?

Answer: No. When light is absorbed, some electrons gain energy. The total energy in the system is increasing.

Example: Consider a semiconductor with a constant voltage applied across it. Is this system in thermal equilibrium?

Answer: No. The electric field accelerates the electrons and holes, causing current flow and increasing the overall energy of the system. This system may be in steady state, but it is not in thermal equilibrium.

Example: Consider a semiconductor coated with an insulator. A voltage is applied to the insulator. No current can flow because of the insulator. An electric field exists within the insulator, and within the semiconductor. Let sufficient time pass such that the charges redistribute themselves within the semiconductor and no more net motion due to the electric field occurs. Is this system in thermal equilibrium?

Answer: No. The electrons near the positive electrode are at a higher potential energy than the electrons elsewhere. The energy distribution is not uniform. This system is in steady state, but it is not in thermal equilibrium.

Example: A candle is placed under one end of a semiconductor. Is the semiconductor in equilibrium?

Answer: No. There is a thermal gradient. The portion of the semiconductor under the candle is hotter than the rest of the semiconductor. In fact, temperature is a measure of the average energy of a system. If the temperature varies, then the average energy must vary, and the system is not in thermal equilibrium.

4.5   Fermi Level in Thermal Equilibrium

When a system is in thermal equilibrium, the average energy at any position (any x) is constant. **The Fermi level is a parameter that is directly related to the average energy of a system**. Therefore, in equilibrium, the Fermi level is constant. Consider Figure 4.1. In Figure 4.1a, the semiconductor is uniformly doped and the energy band diagram shows that the Fermi level is flat as a function of x. In Figure

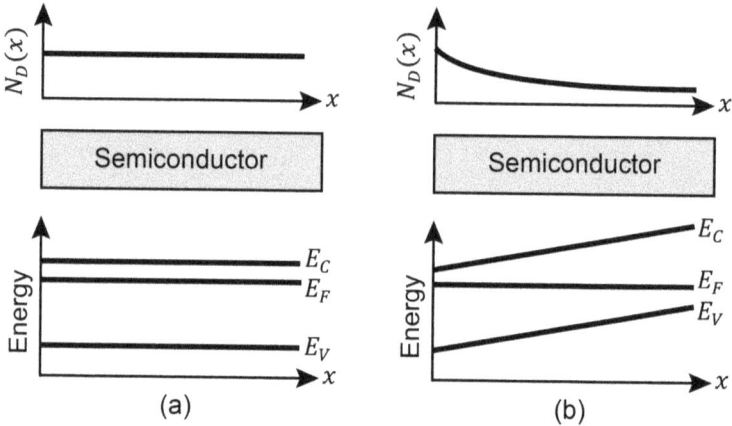

Figure 4.1: Energy band diagrams in thermal equilibrium. (a) A uniform doping concentration results in a flat energy band diagram. (b) In a semiconductor with a non-uniform doping concentration, the Fermi level is still flat, but the conduction band and valence band vary with the doping concentration.

4.1b, the doping concentration of the semiconductor is non-uniform. As long as the semiconductor is in equilibrium, the Fermi level must be constant (flat). On the left, the donor doping concentration is much larger than on the right, and thus E_C is closer to E_F on the left. On the right, the doping concentration is much lower, and thus E_F is closer to the middle of the energy band diagram. Thus, E_C and E_V may vary with respect to E_F throughout the semiconductor in thermal equilibrium, but the average energy, E_F, must be constant in thermal equilibrium.

In thermal equilibrium, the Fermi level is always flat.

Let us look at Figure 4.1b in more detail. Let's start with the electron concentration. The electron concentration is greater on the left because the doping concentration is greater on the left. Now look at the electron energy in the conduction band. Since the electrons are located at the bottom of the conduction band, the electron energy is greater on the right than on the left. Thus, on the left there are a greater number of electrons, but each electron is at a lower energy. On the right there are fewer electrons, but each one is at a larger energy. Including the hole energies, we find that the average energy on the left is the same as the average

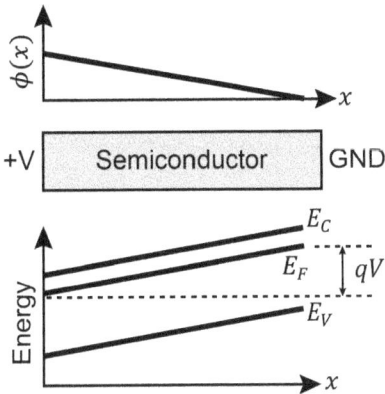

Figure 4.2: When a voltage is applied to a uniformly doped semiconductor, the energy bands (bottom) move in the opposite direction of the electric potential (top).

energy on the right. This is what the constant Fermi level is showing us. If the Fermi level is constant, the average energy is constant. If the Fermi level is not constant, then the average energy is not constant.

4.6 Fermi Level, Energy, and Voltage

The Fermi level is intimately tied to an applied voltage because both the applied voltage and Fermi level are average measures. Let's say a positive voltage is applied to one end of a uniformly doped semiconductor, and the other end is grounded. The electrons at the grounded end will be at a higher energy because the electrons are attracted to the positive voltage. This is illustrated in Figure 4.2. The Fermi level is related to the average energy of ALL the electrons in the semiconductor. This shows an important fact: **The Fermi level and the applied voltage move in opposite directions.** The conduction band and the valence band just went along for the ride. The distance between the conduction band and Fermi level did not change due to the applied voltage, nor did the distance between the valence band and the Fermi level change due to the applied voltage. Figure 4.2 shows this. In Equation form:

$$\Delta E_F = -q\Delta V_{app} \qquad (4.2)$$

A good rule to remember is: **A positive applied voltage lowers the Fermi level.**

The applied voltage only determines the position of the Fermi level at the points where the voltage is applied. The Fermi level doesn't necessarily have a constant slope between the electrodes, although it does in Figure 4.2 because the semiconductor is uniformly doped.

4.7   Review of Relevant Electrostatics

A review of electrostatics is useful because we will, on multiple occasions, work a problem by (1) starting with a charge distribution, (2) integrating once to get the electric field, and (3) integrating a second time to get the electric potential.

4.7.1   Electric Field

An electric field exists in all space due to the charges within that space. Electric field lines originate at positive charges and terminate at negative charges. Mathematically, the electric field can be found from **Gauss's Law**:

$$\nabla \cdot \mathcal{E} = \frac{\rho}{\epsilon_r \epsilon_0} \tag{4.3}$$

where $\mathcal{E}$ is the electric field, ρ is the volume charge density, and the relative dielectric constant and the dielectric constant of free space are ϵ_r and ϵ_0. In general, the electric field, volume charge density, and dielectric constant may all be a function of position (x, y, and z). The electric field is a vector, but can be represented as a scalar if we work in one dimension. A positive electric field points to the right (+x), and a negative electric field points to the left (-x). In 1-D, Gauss's law becomes:

$$\frac{d\mathcal{E}}{dx} = \frac{\rho(x)}{\epsilon_r \epsilon_0} \tag{4.4}$$

At an interface between two materials, the boundary conditions are:

$$\epsilon_1 \mathcal{E}_{1x} = \epsilon_2 \mathcal{E}_{2x} \tag{4.5}$$

where the subscript 'x' was used to clarify that this equation applies for the 1-D case and the electric field is normal to the interface.

4.7.2   Electric Potential

The electric potential is related to the electric field by the following equation: ***

$$\mathcal{E} = -\nabla\phi \qquad (4.6)$$

The electric potential is a scalar and a function of position (x,y,z). In one dimension, this equation becomes:

$$\mathcal{E} = -\frac{d\phi}{dx} \qquad (4.7)$$

Combining Equations 4.4 and 4.7 results in **Poisson's Equation**, relating the electric potential to the volume charge density:

$$\frac{d^2\phi}{dx^2} = -\frac{d\mathcal{E}}{dx} = -\frac{\rho(x)}{\epsilon_r\epsilon_0} \qquad (4.8)$$

Poisson's Equation will be used extensively when studying semiconductor devices such as the MOSFET and diode.

The electric field can be re-written in terms of the conduction band E_C (or E_V or E_i) from Equation 4.7 and using $\Delta E_C = -q\Delta\phi$:

$$\mathcal{E} = -\frac{d\phi}{dx} = \frac{1}{q}\frac{dE_C}{dx} \qquad (4.9)$$

This equation shows that the electric field is always pointed in the direction of increasing conduction band energy.

4.8   Fermi Level, Electric Field, Electric Potential

Let's look at the non-uniform doping distribution used in Figure 4.1b again. In Figure 4.3a is the same non-uniform doping concentration, but the electric potential is also plotted as a function of x. The conduction band is a mirror image of the electric potential. This is always true. If we work in electron-volts and volts and use q equal to "1", the numeric change in the conduction band is equal to the numeric change in the electric potential, although they are different units. In the figure we have to keep the "q" even though its value may be "1" to maintain the correct units.

Here is the crazy thing about Fig. 4.3b. Both sides of the semiconductor are at ground. That is, there is no voltage applied to the semiconductor. And yet, the electric potential changes, and there is an electric field within the semiconductor... with NO APPLIED VOLTAGE!

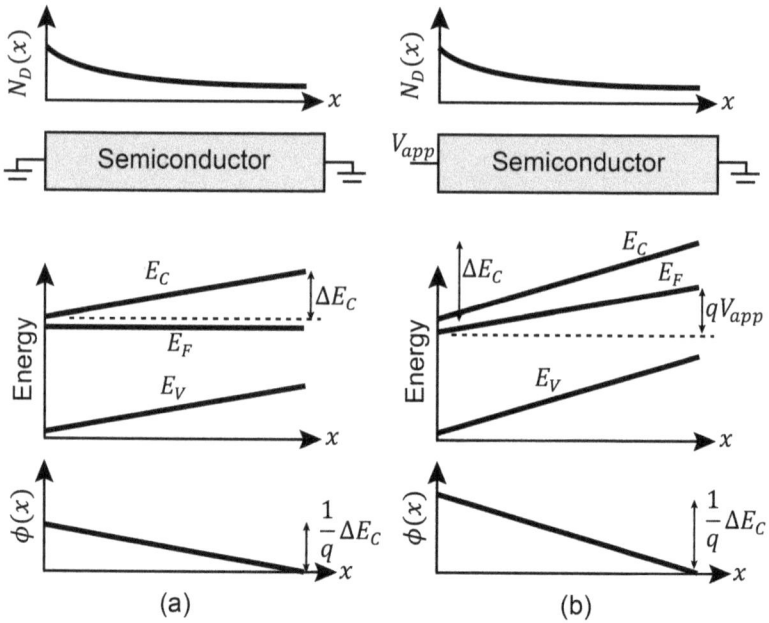

Figure 4.3: Energy band diagrams and electric potential for a non-uniformly doped semiconductor. (a) Zero voltage applied. (b) A positive voltage is applied to the left side.

In Figure 4.3b, a voltage is applied to the semiconductor. An exact solution will require the simultaneous solution of Poisson's Equation, the Continuity Equation (Chapter 6), and the Drift-Diffusion Current Density Equations (Chapter 6), which must be performed numerically. A very close approximation to the solution may be found by assuming the Fermi level varies linearly, and $E_C - E_F$ may be calculated from the doping concentration. The electric potential is found from the mirror image of the conduction band. In this example you can see that the change in electric potential is larger than the applied voltage. Some other points: (1) the band gap does not change, (2) the electric field is pointed to the right, going from high potential to low potential, and (3) the electric field is pointed to the right, in the direction of increasing E_C in agreement with Equation 4.9.

4.9 Volume Charge Density in a Semiconductor

The volume charge density is calculated by summing up all the positive charges per volume and subtracting the negative charges per volume.

We will start by assuming that all the protons cancel with all the electrons in the atomic orbitals up to a *full* valence band. Thus, the starting point is zero volume charge density. The word *full* is key here, as we then have to find all the other charges that deviate from this starting point. Let us consider the effect of holes, electrons, donor atoms, and acceptor atoms:

Holes, $p(x)$: The concentration of holes must be counted as deviating from our starting case because we assumed the valence band was full. Any electron not in the valence band (a hole) does not cancel the charges due to protons, so there is a net positive charge for every hole.

Electrons, $n(x)$: Any electron in the conduction band was not counted, because it is in an atomic orbital higher than the valence band. Thus, each electron in the conduction band contributes a negative charge.

Donor atoms, N_D^+: The electron from each donor atom leaves the atom. This electron is counted elsewhere, such as in the conduction band, or by filling in a hole. There remains one proton that is not balanced by an electron in the full valence band. Thus, there is a net positive charge that has not been accounted for and must be included when calculating the volume charge density.

Acceptor atoms, N_A^-: Each acceptor atom attracts an electron from elsewhere. Since this electron is not accounted for elsewhere, such as in the conduction band or valence band, an additional negative charge must be counted for each acceptor atom when calculating the volume charge density.

The volume charge density is found by summing up all these charges. The units are Coulombs per volume, or C/cm^3.

$$\rho(x) = q[p(x) - n(x) + N_D^+(x) - N_A^-(x)] \qquad (4.10)$$

Looking at Equation 4.10, we see that there are two types of charges in a semiconductor: **mobile charges** and **fixed charges**. The **mobile charges** are the electrons and holes, and the mobile charges can carry current. The **fixed charges** cannot move as they are due to the atoms in a solid. Fixed charges help determine the overall charge density in a semiconductor, but they cannot carry current.

Example 4.7: Consider a uniformly doped semiconductor in thermal equilibrium. What is the volume charge density?

Answer: The total number of protons must equal the total number of electrons. The charge density therefore must be:

$$\rho(x) = 0$$

This equation is actually a very good starting point for most semiconductor problems, and we did not need to use Equation 4.10 to find this answer.

Example 4.8: Consider a uniformly doped semiconductor with a small voltage applied. Current flows through the semiconductor. What is the volume charge density?

Answer: With no current flow, the volume charge density is zero. When current flows, one electron leaves the semiconductor at one end, but a compensating electron enters from the other side. Therefore, the volume charge density is still zero.

In both examples, the volume charge density is zero. When do we need to use Equation 4.10? We will need Equation 4.10 when solving a non-uniform semiconductor. Portions of the semiconductor device may have different volume charge densities. However, in all cases, the net charge will be zero. That is, if there is a region with a negative charge, there must be a region with a compensating positive charge.

4.10 Summary of Key Points

- The word **potential** is ambiguous. You must refer to the **potential energy**, or the **electric potential**.
- An electron-volt (eV) is a measure of energy, not voltage.

- The Fermi level is related to the applied voltage. The Fermi level and applied voltage are both related to the average energy of the system.
- The conduction band is related to the electric potential. These are both related to the individual energy of the electrons.
- The electric field points from a positive charge to a negative charge.
- The Fermi level is flat in thermal equilibrium.
- The applied voltage determines the Fermi level at the electrodes only. The Fermi level within the semiconductor must be calculated.
- When a positive voltage is applied to an electrode, the Fermi level is reduced at that location.
- An assumption is made throughout this book that the Fermi-Dirac and Maxwell-Boltzmann distributions are valid even when we are not in thermal equilibrium. This is generally a very good approximation.
- Poisson's equation relates the electric potential, electric field, and charge density.
- The volume charge density is made up of the fixed dopant atoms (acceptor and donor atoms) and the mobile charge carriers (electrons and holes).
- Dopant atoms cannot move. Therefore, the charges associated with the dopant atoms cannot carry current.
- The mobile charge carriers are the electrons and the holes. These can move throughout a semiconductor, and therefore electrons and holes can carry current.
- The overall charge of a semiconductor must be zero.

4.11 Problems

1. Consider the following band diagrams. Putting a probe on either end, what voltage will you measure using a voltmeter for each case?

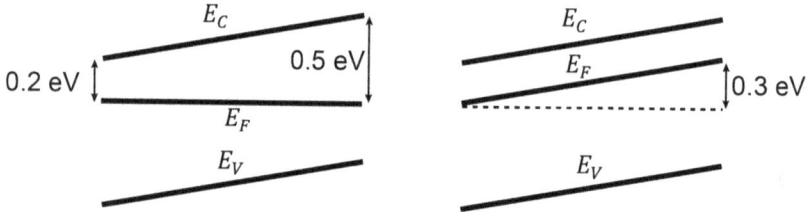

2. Figure 4.1b shows an example of a semiconductor with a non-uniform doping profile that results in a linear increase in E_C relative to E_F. Let $x = 0$ represent the left side of the semiconductor.

   a. Using the equation of a line $(y = mx + b)$, write an equation that represents $E_C - E_F$ as a function of x to obtain the energy band profile shown in Figure 4.1b.

   b. Now find an equation for the doping concentration as a function of x using Equation 3.5. That is, find $N_D(x)$.

   c. Using Equation 4.9, find the electric field as a function of x for this non-uniformly doped semiconductor. Is the electric field a constant?

3. In later chapters, we will see how to change the charge distribution in a semiconductor. For now, let us assume that all the electrons and holes in a semiconductor are somehow removed. That is, $n(x) = p(x) = 0$. Further, assume that the semiconductor is n-type ($N_A = 0$) and uniformly doped.

   a. Write an equation for the volume charge density, ρ in terms of N_D.

   b. Using Equation 4.4, find the electric field as a function of x. Use the boundary condition that the electric field is zero at $x = 0$.

   c. Using Equation 4.7, find the electric potential as a function of x. Use the boundary condition that the electric potential is zero at $x = 0$.

4. Consider the following charge distribution in a semiconductor.
   a. What is the electric field as a function of x, using the boundary condition that the electric field is zero at x=0?
   b. What is the maximum electric field?
   c. What is the electric potential as a function of x, using the boundary condition that the electric potential is zero at x=0?
   d. What is the maximum electric potential?
   e. Draw the energy band diagram for this semiconductor.

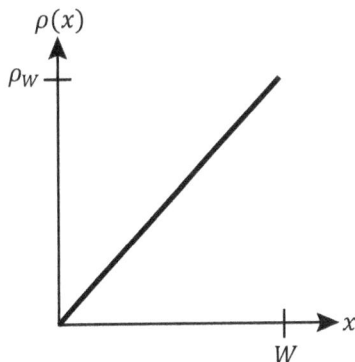

5   Long Channel MOSFETs

A good starting point for understanding transistors is the long-channel **MOSFET**. The long-channel MOSFET is not used in practice, as they are rather large and expensive, but they are a fundamental component of the small, modern MOSFET. They also illustrate some of the key principles used in all types of transistors.

A diagram of the device is shown in Figure 5.1. On the left is a 3-D diagram of MOSFET. The action occurs within the silicon, so a cross-section of the device is shown on the right. There are four metal electrodes called the **gate**, **drain**, **source**, and **body (substrate)**. These electrodes connect to the MOSFET. The names body and substrate are both used, but the letter B is always used when labeling it to avoid confusion with the source. The source and drain are identical, but it will

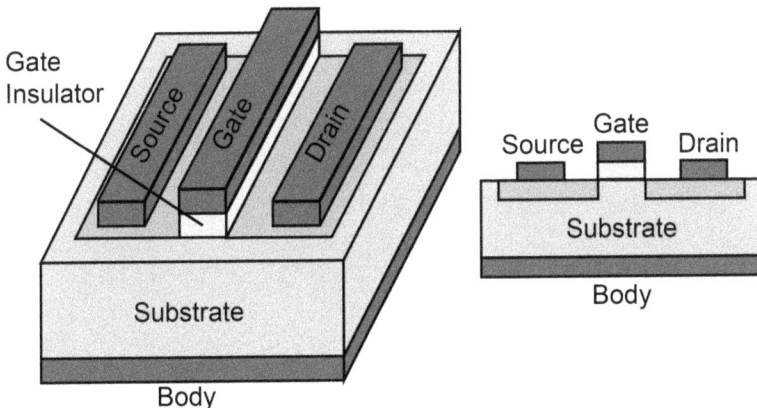

Figure 5.1. (left) A 3-D representation of a MOSFET. The source and drain electrodes are connected to the substrate. The gate electrode is insulated from the substrate by an insulator. (right) A cross-section of the MOSFET showing the source and drain regions within the substrate.

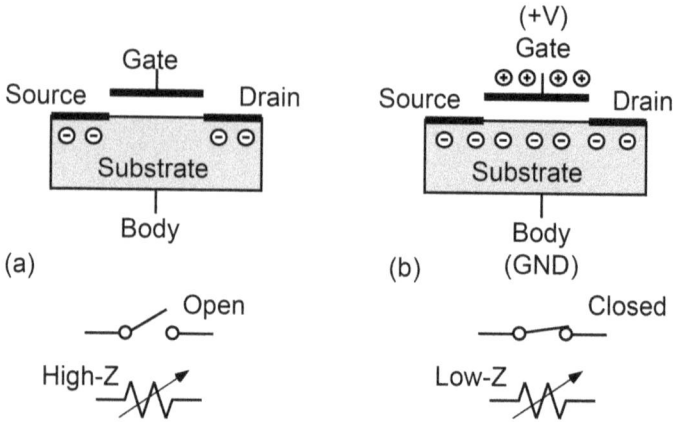

Figure 5.2: Illustrative example showing how a MOSFET operates. (a) With no voltage applied, there is no electrical connection between the drain and source. (b) When a positive voltage is applied, the positive charges on the gate attract negative charges to the surface of the semiconductor, connecting the source and drain.

be necessary to label one the source and one the drain to clearly indicate the voltages applied. In a MOSFET, all the current flows between the source and drain. The source and drain are regions within the silicon substrate that are doped with atoms different from the silicon substrate. The gate is made of a conductive material, such as a metal or heavily doped silicon. Heavily doped silicon has traditionally been used for manufacturing reasons, but metal gates are common for high performance MOSFETs. The gate is special as it is surrounded by an insulator, and thus no current flows from the gate to the rest of the transistor. The insulator between the gate electrode and the substrate is called a **gate insulator**.

5.1 Principle of Operation

The MOSFET can be thought of as two interacting devices: a variable resistor and a capacitor. A capacitor exists between the semiconductor substrate and the gate, as shown in Figure 5.2. The region of semiconductor between the source and drain may be considered to be a variable resistor. The source and drain will be formed using n doped silicon, which is full of electrons. With no voltage applied to the gate (Figure 5.2a), the electrons in the source and drain are isolated from each other, and no current can flow. With a positive voltage is applied to the gate (Figure 5.2b), the gate-substrate capacitor charges up with positive charges on the gate and negative charges on the substrate. These negative charges (electrons) connect the source to the drain. Using a switch analogy, the switch is open with no gate voltage applied, and the switch is closed with a positive gate voltage applied. Using a variable resistor analogy, the resistor between the source and drain has a high impedance with no voltage applied to the gate, and the resistance is small with a positive gate voltage applied. The resistor value depends on the number of electrons connecting the source and drain, and thus depends on the gate voltage. Thus, the MOSFET may be thought of as a voltage-controlled resistor.

The region between the source and drain, where the charges accumulate at the semiconductor surface, is called the **channel**. If the surface of the channel changes from p-type to n-type, as shown in Figure 5.2b, we say that an **inversion layer** formed.

Figure 5.3 shows the cross-section of two types of MOSFETs: the **p-channel MOSFET** and the **n-channel MOSFET**. The phrase 'p-channel MOSFET' is long and is usually shortened to **PMOS**. Similarly, an n-channel MOSFET is called an **NMOS**.

PMOS: In the PMOS transistor, the source and drain electrodes are both connected to a p-type semiconductor region. The region separating the source and drain must be n-type to prevent current flow when there is no inversion layer. When a negative voltage is applied to the gate, negative charges accumulate on the gate and positive charges accumulate in the substrate. Holes will accumulate in the substrate, changing it from n-type to p-type, and connecting the p-type regions of the source and drain. This is an amazing feature of semiconductors, because a region that is doped n-type can be turned into a p-type region just by the application of a voltage.

NMOS: In the NMOS transistor, the source and drain are both connected to an n-type semiconductor. The region separating the source and drain must be p-type to prevent current flow when there is no inversion layer. When a positive voltage is applied to the gate, positive charges accumulate on the gate and negative charges accumulate in the

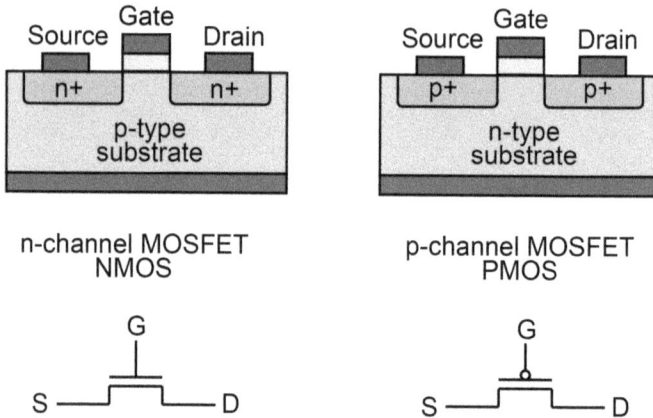

Figure 5.3: Cross-sections of an NMOS and PMOS transistor. The 'n' represents a region n-type, 'p' represents a region that is p-type, and a '+' sign indicates that the doping concentration is very large. Below the cross-sections are the symbols used to represent the NMOS and PMOS transistor.

substrate. Electrons will accumulate in the substrate, changing it from p-type to n-type, and connecting the n-type regions of the source and drain.

Warning! A common exam question is to ask about the relation between NMOS, PMOS, n-channel MOSFET, p-channel MOSFET, n-type substrate, and p-type substrate.

n-channel MOSFET	p-channel MOSFET
NMOS	PMOS
p-type substrate	n-type substrate

5.2   Digital CMOS Logic

A great feature of the MOSFET is that zero current flows into the gate. This minimizes power consumption. In digital circuits, the static power consumption can be zero! That is, power is only consumed when the digital logic changes state. This feature has made the MOSFET the most common transistor choice for digital logic. Billions of transistors can be put on a single computer chip, and the total power consumption can be less than 1 W.

NAND
Gate

Inverter

+5 V

Truth Table

A B	X Y
0 0	1 0
0 1	1 0
1 0	1 0
1 1	0 1

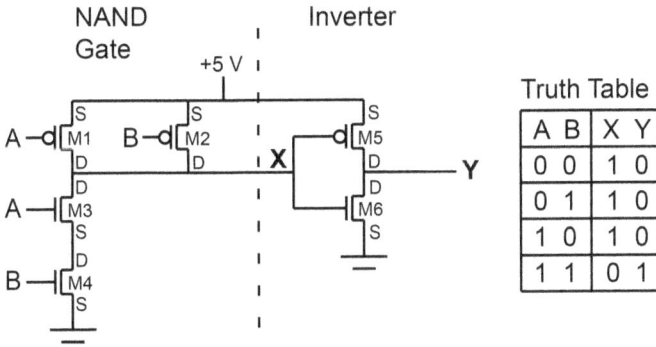

Figure 5.4: A NAND gate connected to an inverter. A truth table is shown to the right.

An example of a simple digital circuit, a 2-input NAND gate connected to an inverter to form an AND gate, is shown in Figure 5.4. The NAND gate consists of two NMOS and two PMOS transistors, and the inverter contains a single NMOS and PMOS transistor. Inputs A and B will be at either 0 V or 5 V. Consider transistor M4. If B is high (+5 V), then transistor M4 will be turned on because it is an NMOS transistor. If B is low (0 V), then transistor M4 will be turned off. Now consider transistor M1, a PMOS transistor. The PMOS transistor is turned on with a negative voltage, but voltages are always relative. Thus, if 'A' is ground, then the gate is more negative than the source and the transistor turns on. If A is high (+5 V), then the gate and source are at the same voltage and the transistor if off. Thus, the output of the NAND gate (X) will be connected to either the +5 V line or ground, depending on the values of A and B. The circuit is designed such that X cannot be connected to both +5 V and ground at the same time. The output Y is also connected to either +5 V or ground. The truth table is shown in Figure 5.4, where a '0' represents zero voltage and a '1' represents +5 V.

5.3 NMOS Energy Band Diagrams

In this section we will look at the semiconductor region of a MOSFET underneath the gate. The energy band diagram of the semiconductor will be analyzed for four cases: accumulation, depletion, weak inversion, and strong inversion. Figure 5.5a shows a cutline of the region of interest. We will use 'y' to indicate this direction since it is a

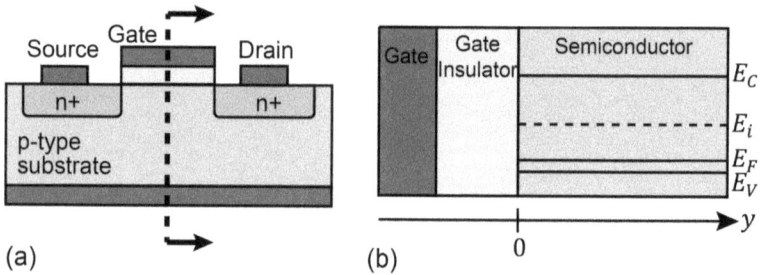

Figure 5.5: (a) A schematic of an NMOS with a cutline showing the location of the y-axis. (b) The y-axis with the gate, gate insulator, and semiconductor regions labeled. The energy bands are shown within the semiconductor region.

vertical cut through the MOSFET. The top of the semiconductor is located at $y = 0$. The gate electrode and gate insulator are located at negative values of y. This structure looks like a capacitor: the gate electrode, gate insulator, and semiconductor form the capacitor. Figure 5.5b shows an energy band diagram of an NMOS with a p-type substrate. Currently the Fermi level is below E_i because the semiconductor is doped p-type, but we will soon see that we can change this by applying a voltage to the gate.

The MOSFET mode is determined by the gate voltage, but voltage is relative, so we must specify a reference point. It is convention to treat the source as the reference and derive all the equations using this convention. Thus, the gate voltage of interest is the gate voltage relative to the source voltage. Mathematically, we write:

$$V_{GS} = V_G - V_S$$

5.3.1 NMOS in Accumulation

The best starting point for drawing an energy band diagram is to consider the case called **accumulation**. We will calculate a voltage called the **Flat Band Voltage**, V_{FB}, later in this chapter. If a gate voltage more negative than the Flat Band Voltage is applied to the gate ($V_{GS} \leq V_{FB}$), negative charges accumulate on the gate electrode and positive charges accumulate on the semiconductor surface; hence the name accumulation. The positive charges stay near the semiconductor

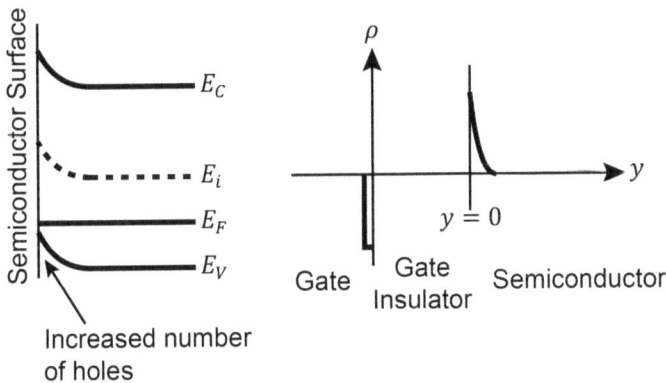

Figure 5.6: Semiconductor band diagram and charge distribution for an NMOS in accumulation; $V_{GS} \leq V_{FB}$.

surface because they are attracted to the negative charges on the gate. Figure 5.6 shows the energy band diagram and the volume charge density distribution for accumulation. There are negative charges on the gate, and positive charges on the semiconductor surface, with zero charges in the gate insulator. The total negative charge on the gate must equal the total positive charge in the semiconductor. The Fermi level at the semiconductor surface is near E_V because there are more holes (positive charges) here. Notice that the band gap does not change. The charges in the semiconductor are all within a few nm of the surface. Although possible, it is difficult to get the Fermi level to cross the valence band, and our equations will break down because the Bose-Einstein approximation fails. Therefore, in this book, we will use the approximation that **the valence band does not cross the Fermi level**.

5.3.2   NMOS in Depletion and Weak Inversion

If the gate voltage is more positive than the flat band voltage ($V_{GS} \geq V_{FB}$), then the energy bands will bend in the opposite direction from accumulation. The true limits for depletion and weak inversion are $V_T \geq V_{GS} \geq V_{FB}$, where V_T will be defined in the next section. The conduction band and valence band will move down at the semiconductor surface, moving the Fermi level away from the valence band. This will reduce the number of holes near the surface. In fact, it is possible for E_i to cross E_F. Figure 5.7 shows two band diagrams. We call it **depletion** when E_i does not cross E_F, and we call it **weak**

inversion when E_i crosses E_F. The meaning behind these terms will become clear from the charge distribution. The volume charge density, derived in Chapter 4, is repeated here:

$$\rho(x) = q[p(x) - n(x) + N_D^+(x) - N_A^-(x)] \tag{5.1}$$

We can guess that there will be negative charges in the semiconductor because the positive gate voltage will attract negative charges to the semiconductor surface. The negative charges do not need to be electrons! Let's examine this exciting case in more detail.

Let us define a location, labeled $y = W_D$, where the energy bands become approximately the same as their equilibrium values. This location is shown in Figure 5.7. W_D represents the depletion layer width, which will be explained in a bit. The volume charge density in the region from $y = 0$ to $y = W_D$ can be calculated. Let us rewrite Equations 3.12 and 3.13 for n and p for the non-equilibrium condition:

$$n(y) = n_i \exp\left(\frac{E_{Fi}(y)}{kT}\right) \tag{5.2}$$

$$p(y) = n_i \exp\left(-\frac{E_{Fi}(y)}{kT}\right) \tag{5.3}$$

Since the semiconductor is p-type, there are no donor atoms. Assuming that the acceptor atoms are uniformly distributed, we get:

$$N_D^+ = 0 \tag{5.4}$$

$$N_A^- = \text{constant} \tag{5.5}$$

We can now find the charge density from Equation 5.1:

$$\rho = q\left[n_i \exp\left(-\frac{E_{Fi}(y)}{k_B T}\right) - n_i \exp\left(\frac{E_{Fi}(y)}{k_B T}\right) + 0 - N_A^-\right] \tag{5.6}$$

Notice that E_F is close to the middle of the band gap in this region, and therefore $p(y)$ and $n(y)$ are both significantly smaller than N_A^-. Thus,

$$\rho \approx q(\text{small} - \text{small} + 0 - N_A^-) \tag{5.7}$$

$$\rho = -qN_A^- \quad \text{(NMOS)} \tag{5.8}$$

Depletion

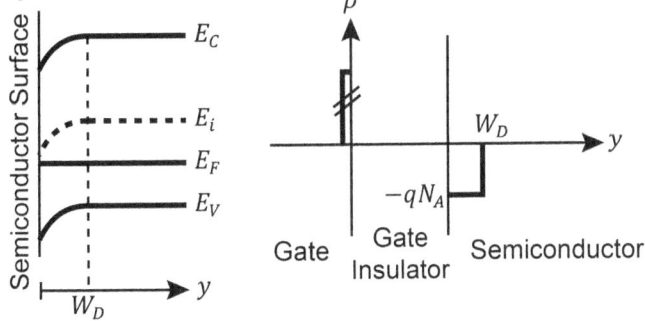

Weak Inversion

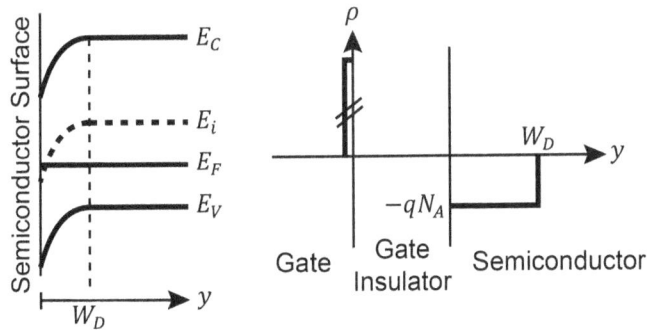

Figure 5.7. Band diagram and volume charge density for an NMOS in depletion and weak inversion; $V_{FB} \leq V_{GS} \leq V_T$. The volume charge density in the gate is shown with hash marks on the vertical scale because the area under the curve should equal the area under the curve of the semiconductor. Note that E_F does not cross E_i in depletion, but does cross E_i in weak inversion.

$$\rho = qN_D^+ \quad \text{(PMOS)} \tag{5.9}$$

where the PMOS result is given for completeness. This is an amazing result. It says that the volume charge density is negative and constant from $y = 0$ to $y = W_D$. The negative charge was expected because we were trying to get electrons at the silicon surface to turn on the MOSFET. However, we did NOT get a lot of electrons here. The charges are due to the ionized acceptor atoms left in place after the holes were pushed away from the surface. Keep in mind that the atoms do not move. This region is called a **depletion layer** because the **mobile**

carriers (electrons and holes) are small compared to the charge due to the **fixed charges** (acceptor and donor atoms). The mobile charges are depleted from this region. It will take a larger voltage to get a lot of electrons in the channel.

For the region deeper than $y = W_D$, the volume charge density is zero because the semiconductor approximates the equilibrium condition. Figure 5.7 shows a plot of the volume charge density as a function of depth. The only difference between the depletion case and the weak inversion case is the depth of the depletion region. Included on this plot is the gate insulator (no charge) and the gate electrode. The magnitude of the charge on the gate must be exactly equal to the charge in the semiconductor because the gate-insulator-semiconductor acts like a capacitor.

5.3.3   NMOS in Strong Inversion

If the gate voltage is continually increased, taking the MOSFET past depletion and into weak inversion, the electron concentration near the surface will continually increase. Eventually, the volume charge density due to the electrons will equal the volume charge density due to the acceptor atoms. The gate voltage at which this occurs is called the **threshold voltage**, V_T. For gate voltages greater than the threshold voltage ($V_{GS} \geq V_T$), the NMOS is in **strong inversion**.

In strong inversion, the second term in Equation 5.6, which represents the electron concentration, is no longer small and must be kept. In fact, as the gate voltage increases, the second term will become large, and the volume charge density at the surface will be:

$$\rho \approx q(\text{small} - \text{large} + 0 - N_A^-) \tag{5.10}$$

The band diagram and volume charge density are plotted in Figure 5.8. There is a layer of electrons at the surface of the semiconductor and the depletion region underneath it. Effectively, the width of the depletion region has reached its maximum value and an increase in the gate voltage causes an increase in the electron concentration at the semiconductor surface. This depletion region width is labeled W_{Dm}, where the 'm' represents maximum. The layer of electrons at the semiconductor surface is called the **inversion layer**. The name

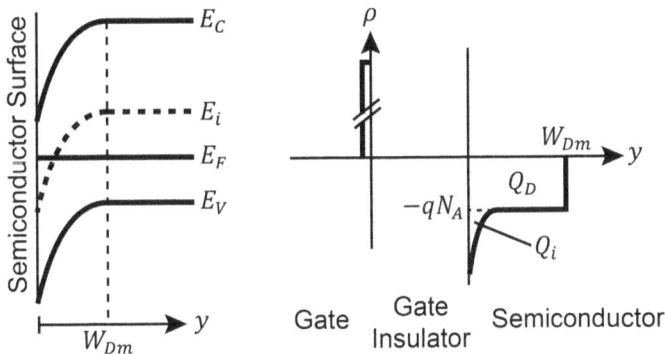

Figure 5.8: The band diagram and the volume charge density of an NMOS in strong inversion; $V_{GS} \geq V_T$. The surface charge in the semiconductor (Q_S) is made up of the depletion charge (Q_D) and the inversion charge (Q_i).

inversion describes the fact that we started with p-type silicon filled with holes, and we inverted it to be filled with electrons.

The electrons at the surface of the semiconductor connect the n-type source with the n-type drain. Therefore, when $V_{GS} \geq V_T$ and the NMOS is in strong inversion, the NMOS is turned on and current can flow from the drain to the source. In accumulation, weak inversion, and depletion, the MOSFET is turned off and no current flows.

The charges in the semiconductor consist of the charges that make up the depletion region called the **depletion charge** (Q_D), and the electrons at the semiconductor surface, called the **inversion charge** (Q_i).

5.4   PMOS Energy Band Diagrams

The PMOS energy band diagram looks similar to a mirrored NMOS band diagram, except the Fermi level starts above E_i because the semiconductor is doped n-type. The volume charge distribution is also a mirror image of the NMOS volume charge distribution.

Accumulation: The energy bands bend down in accumulation, causing a large number of electrons to accumulate near the surface. For the PMOS to be in accumulation, $V_{GS} \geq V_{FB}$ because a positive voltage causes the energy bands to move down.

Depletion and Weak Inversion: The energy bands bend up in depletion, creating a depletion region as the electron concentration decreases and the hole concentration increases. The hole concentration does not increase sufficiently to overcome the donor dopant concentration. For a PMOS, depletion and weak inversion occur for $V_T \leq V_{GS} \leq V_{FB}$.

Strong Inversion: The PMOS transistor is turned on for $V_{GS} \leq V_T$. Here, the energy bands bend up at the surface such that the hole concentration is comparable to or greater than the donor dopant concentration. The holes at the surface create an inversion layer that connect the source and drain. Beneath the inversion layer is a depletion region.

5.5   Threshold Voltage

The threshold voltage is one of the most important parameters of a MOSFET. In this section, we will look at the properties of the semiconductor substrate and the gate insulator to determine the threshold voltage. The semiconductor substrate will be assumed to be silicon for this chapter because silicon is the most commonly used material. In a later chapter when we discuss modern MOSFETs, we will look at other materials that could be used, and find out why silicon is used in well over 99% of all MOSFETs.

The **threshold voltage** (V_T) is defined as the voltage applied to the gate that causes the semiconductor surface to be p-type just as much as it was n-type (PMOS), or n-type just as much as it was p-type (NMOS). For example, if the substrate is doped such that $N_A = 10^{16} \text{cm}^{-3}$, then the semiconductor surface must have an electron concentration equal to 10^{16} cm^{-3} when $V_{GS} = V_T$.

Shown in Figure 5.9 is an energy band diagram in the silicon at the threshold condition, $V_{GS} = V_T$, using the NMOS transistor as an example. This band diagram will look similar to Figure 5.8, but the distance that the energy bands bend will be very specific to the threshold condition. The electric potential is also shown in Figure 5.9. The

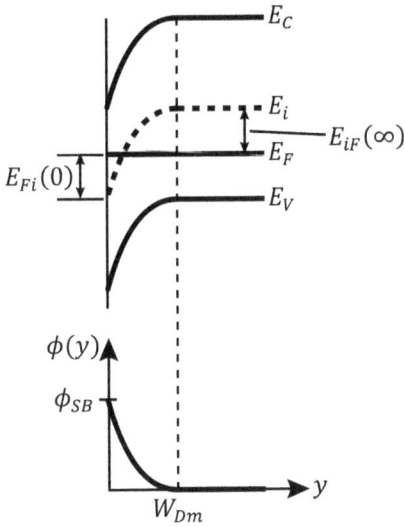

Figure 5.9: The energy bands and the electric potential are shown for an NMOS at the threshold condition ($V_{GS} = V_T$).

electron concentration must equal N_A, as this is the definition of threshold. That is,

$$n(0) = N_A \qquad (5.11)$$

The electric potential between a point deep in the substrate and semiconductor surface will now be calculated. The symbol to be used is ϕ_{SB} where the 'S' represents surface and the 'B' represents body. Unfortunately, the subscript 'S' has to do double duty as it is also used to represent the source, but hopefully the context will make it clear what 'S' refers to. ϕ_{SB} is indicated on Figure 5.9 and is equal to:

$$\phi_{SB} = -\frac{1}{q}\left(E_C(0) - E_C(\infty)\right) \qquad (5.12)$$

where the ∞ symbol is used to indicate a point deep in the substrate. Written in terms of the middle of the band gap, ϕ_{SB} becomes:

$$
\begin{aligned}
\phi_{SB} &= -\frac{1}{q}\left(E_i(0) - E_i(\infty)\right) \\
&= -\frac{1}{q}\left(E_i(0) - E_F - E_i(\infty) + E_F\right) \qquad (5.13) \\
&= \frac{1}{q}\left(E_{Fi}(0) + E_{iF}(\infty)\right)
\end{aligned}
$$

Rewriting Equations 3.12 and 3.13:

$$E_{Fi}(y = 0) = k_B T \ln \frac{n(0)}{n_i} \tag{5.14}$$

$$E_{iF}(y = \infty) = k_B T \ln \frac{p(\infty)}{n_i} \tag{5.15}$$

Substituting Equations 5.14 and 5.15 into 5.13, and using $n(0) = p(\infty) = N_A$, we obtain:

$$\boxed{\phi_{SB} = 2 \frac{k_B T}{q} \ln \frac{N_A}{n_i}} \quad \text{(NMOS)} \tag{5.16}$$

$$\boxed{\phi_{SB} = -2 \frac{k_B T}{q} \ln \frac{N_D}{n_i}} \quad \text{(PMOS)} \tag{5.17}$$

The maximum depletion width, W_{Dm}, can be found from electrostatics. In this case, we will use Poisson's Equation (Equation 4.8):

$$\frac{d^2 \phi}{dy^2} = -\frac{\rho}{\epsilon_S \epsilon_0} \tag{5.18}$$

This equation needs to be integrated twice to get the potential at the surface. The reference potential will be a location deep in the substrate (body), and the integration will occur from $y = W_{Dm}$ to $y = 0$. The surface potential is then found to be:

$$
\begin{aligned}
\phi_{SB} &= -\frac{1}{\epsilon_S \epsilon_0} \int_{W_{Dm}}^{0} \int_{y}^{0} \rho \, dy \\
&= -\frac{1}{\epsilon_S \epsilon_0} \int_{W_{Dm}}^{0} \int_{y}^{0} (-)q N_A^- \, dy \\
&= -\frac{1}{\epsilon_S \epsilon_0} \int_{W_{Dm}}^{0} q N_A^- y \, dy \\
&= \frac{1}{2 \epsilon_S \epsilon_0} q N_A^- W_{Dm}^2
\end{aligned}
\tag{5.19}
$$

Solving for W_{Dm}:

$$W_{Dm} = \sqrt{\frac{2 \epsilon_S \epsilon_0 \phi_{SB}}{q N_A^-}} \quad \text{(NMOS)} \tag{5.20}$$

$$W_{Dm} = \sqrt{\frac{2\epsilon_s\epsilon_0|\phi_{SB}|}{qN_D^+}} \quad \text{(PMOS)} \tag{5.21}$$

The surface charge density (Q_D) can be found by integrating the volume charge density from $y = 0$ to $y = W_{Dm}$. This is simply the area of a rectangle, as seen in Figure 5.8 without the inversion charge. When integrating over a depletion region, the surface charge density is known as the **depletion charge density**, and has units of C/cm^2.

$$Q_D = -qN_A^-W_{Dm} \quad \text{(NMOS)} \tag{5.22}$$

$$Q_D = qN_D^+W_{Dm} \quad \text{(PMOS)} \tag{5.23}$$

Be careful with the symbols. We are distinguishing between a **volume** charge density, ρ, with units C/cm^3 and a **surface** charge density, using the symbol Q, with units of C/cm^2. A volume charge density may be converted to a surface charge density by integrating over the semiconductor thickness:

$$Q = \int \rho \, dy \tag{5.24}$$

The threshold voltage can now be calculated. The threshold voltage is defined from the viewpoint of the gate electrode. The threshold voltage is found by summing the electric potentials from the substrate up to the gate, plus another factor called a flat-band voltage that will be discussed in the next section. The electric potential from the body to the surface of the silicon is ϕ_{SB}, as shown in Figure 5.10. The electric potential across the gate insulator is found from the equation for the charge on a capacitor: $Q = CV$, where Q and C are charge per area and capacitance per area, respectively.

$$\phi_{GS} = \frac{Q_G}{C_i} = -\frac{Q_D}{C_i} \tag{5.25}$$

where ϕ_{GS} is the voltage across the gate insulator, from the gate to the semiconductor surface, and Q_G is the charge per area on the gate. We have ignored the inversion charge density because, when $V_{GS} = V_T$, the inversion layer has just formed and $Q_i \ll Q_D$. C_i is the capacitance per unit area (F/cm^2) of the gate insulator and is equal to:

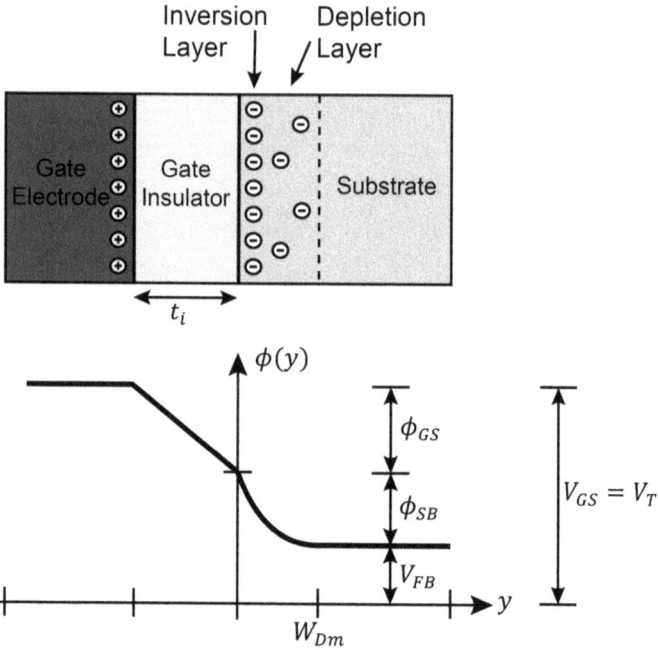

Figure 5.10: Electric potential versus distance for an NMOS at the threshold condition.

$$C_i = \frac{\epsilon_i \epsilon_0}{t_i}$$

(5.26)

where ϵ_i is the relative dielectric constant of the insulator and t_i is the insulator thickness. Using Figure 5.10 as a helpful guide to add up all the electric potentials, the threshold voltage is then:

$$V_T = V_{FB} + \phi_{SB} - \frac{Q_D}{C_i}$$

(5.27)

Note that the second term is positive for NMOS, and negative for PMOS. Also, the third term, the depletion charge density, will have a net positive effect on V_T due to the negative depletion charge in an NMOS transistor, and be negative for a PMOS transistor. Inserting Equation 5.22 for Q_D, we obtain:

$$V_T = V_{FB} + \phi_{SB} + \frac{\sqrt{2\epsilon_S \epsilon_0 q N_A^- \phi_{SB}}}{C_i}$$

(NMOS)   (5.28)

$$V_T = V_{FB} + \phi_{SB} - \frac{\sqrt{2\epsilon_S\epsilon_0 q N_D^+ |\phi_{SB}|}}{C_i} \qquad \text{(PMOS)} \qquad (5.29)$$

> **Concept Review**. When the gate voltage is larger than the threshold voltage, the MOSFET is <u>**ON**</u> and current can flow from the drain to the source. When the gate voltage is smaller than the threshold voltage, the MOSFET is <u>**OFF**</u> and no current flows between the drain and source. In equation form:
>
> $$\text{if } V_{GS} \begin{cases} < V_T \text{ then } I_{DS} = 0 & \textbf{(OFF)} \\ \geq V_T \text{ then } I_{DS} \geq 0 & \textbf{(ON)} \end{cases} \quad \text{(NMOS)}$$
>
> $$\text{if } V_{GS} \begin{cases} > V_T \text{ then } I_{DS} = 0 & \textbf{(OFF)} \\ \leq V_T \text{ then } I_{DS} \leq 0 & \textbf{(ON)} \end{cases} \quad \text{(PMOS)}$$

5.6 Flat-Band Voltage

When calculating the threshold voltage, a term called the flat-band voltage was introduced but not defined. The flat band voltage is the gate voltage required to make the energy bands flat in the semiconductor. Since the electric potential is a mirror image of the energy bands, $\phi_{SB} = 0$ when $V_{GS} = V_{FB}$. In Figure 5.10, we can see that if the gate voltage were equal to the flat-band voltage, there would be no voltage drop across the gate oxide or in the semiconductor, making the energy bands flat. Two effects have been ignored up to now that give rise to the flat band voltage. First, there may be a difference in the work functions of the gate and semiconductor. Second, there may be charges within the gate insulator that we did not take into account.

5.6.1 Work Function Difference

The **work function** of a material is defined as the average energy required to remove an electron from that material. This is equal to the energy required to move an electron from the Fermi level to the vacuum level, E_{vac}. This nomenclature, by convention, does not follow the single subscript rule.

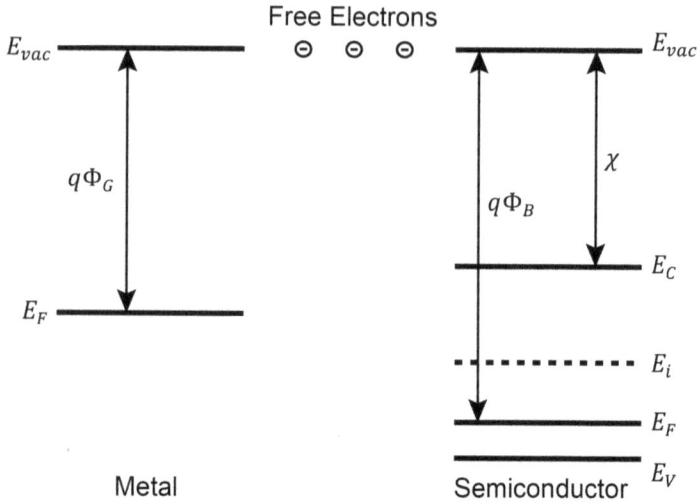

Figure 5.11: Energy band diagram of a metal (left) and a semiconductor (right) that are separated from each other. Some free electrons are also indicated.

Let us start by considering two materials: a metal gate and a semiconductor substrate. The energy band diagram of the two materials is shown in Figure 5.11. In the metal, there is no conduction band or valence band because there is no band gap. The electrons can occupy a continuous range of energies. The average electron energy is at the Fermi level; some electrons have a higher energy and some electrons have a lower energy. There is no valence band, and therefore there are no holes in a metal.

In addition to the Fermi level, there is a vacuum level, indicated as E_{vac}. In Figure 5.11 are shown some free electrons between the metal band diagram and the semiconductor band diagram. These electrons could have come from either the metal or the semiconductor. The amount of energy to remove these electrons will vary for the two materials, but the energy of the free electrons is the same once they are removed from the material. Therefore, it makes sense to **use the energy of the free electrons, or E_{vac}, as a reference energy when drawing band diagrams involving multiple materials.**

The semiconductor is drawn with energy bands for E_{vac}, E_C, E_V, E_F, and E_i. The **electron affinity**, χ, is the distance from the vacuum level to the conduction band, and this is a material property. Values for the electron affinity may be found in Appendix A.

Rules to Follow When Drawing an Energy Band Diagram

1. In equilibrium, the Fermi energy level is constant. It is a straight line through all the materials.

2. With an external voltage applied between two points, the Fermi level between those points has to shift by an amount equal to $\Delta E_F = -qV_{12}$. (Determining the Fermi level between those points may or may not be easy.)

3. The vacuum level is continuous, even across materials. It does not need to be a straight line.

4. Do not draw a valence band or conduction band for a metal.

5. In a metal, the work function is fixed. There is a table of work functions for many metals in Appendix A.

6. In a semiconductor or insulator, the energy difference between the vacuum level and conduction band is fixed. This difference, called the **electron affinity**, is a material constant.

7. The slope of the vacuum level (or conduction band or valence band) is proportional to the electric field. $\mathcal{E} = \frac{1}{q}\frac{dE_C}{dx}$

8. An electric field cannot exist within a metal. All the charges get pushed to the surface. Therefore, the vacuum level and Fermi level do not change within a metal.

9. If there are no charges, such as in an insulator, then E_{vac}, E_C, and E_V will be a straight line (not necessarily horizontal).

10. If there are charges present, such as in a semiconductor, and the energy bands are bending, they will have an arc (not a straight line) due to the termination of the electric field lines.

The work function is defined as the difference between the vacuum energy and the Fermi level (in units of electric potential). Thus, the work function is:

$$\Phi \equiv \frac{1}{q}(E_{vac} - E_F) \qquad (5.30)$$

In our equations we will use a work function in terms of an electric potential instead of energy because it will be more convenient for calculating the flat band voltage. Φ_G is used to represent the work function in the gate. In a metal, the work function is a constant. Some work functions are listed in Table A.2 in Appendix A. The symbol Φ_B is used to represent the work function for the semiconductor (body). The work function is defined the same way as for a metal, but since the location of the Fermi level may be changed by doping, the work function in a semiconductor is not a material property. It must be calculated.

To calculate the work function in a semiconductor, it is convenient use Fig. 5.11b to find the work function as the sum of three terms:

$$\Phi_B \equiv \frac{1}{q}\left\{\chi + \frac{E_G}{2} + (E_i - E_F)\right\} \tag{5.31}$$

Be careful with this equation. While χ and E_G are material properties, $E_i - E_F$ must be calculated based on the doping type and concentration. $E_i - E_F$ may be either positive or negative: it is positive for p-type material and negative for n-type material.

The table on the previous page shows a set of rules to use when drawing energy band diagrams. These rules aren't new; they are simply a restatement of previous facts in one convenient place. When drawing an energy band diagram, I recommend drawing the energy band diagram in three steps:

1. Start with the materials separated. Draw the energy band diagrams for each material far apart, with a constant vacuum level, similar to Figure 5.11.
2. Now draw the energy band diagrams in equilibrium. The first step is to align the Fermi levels. The second step is to draw a continuous vacuum level. Use the Fermi level as a guide. Finally, draw the other energy bands relative to the vacuum level.
3. Finally, apply a voltage to the materials by moving the Fermi level appropriately. Draw the energy band diagrams in the same order as step #2.

I highly recommend that you draw the band diagram in equilibrium first, even if the problem calls for a non-equilibrium condition. You will reduce the number of mistakes significantly!

Example 5.1

Consider an NMOS transistor with an aluminum gate, 10 nm thick SiO$_2$ gate insulator, and a silicon substrate with $N_A = 10^{17}$ cm^{-3}. (a) Draw the energy band diagram for $V_{GS} = 0$ V. (b) What gate voltage will make the semiconductor energy bands flat? (c) Draw the band diagram with flat energy bands.

Solution

From Appendix A, we find the gate work function to be 4.1 V, the SiO$_2$ electron affinity to be 0.9 eV, and the silicon electron affinity to be 4.05 eV. To get the semiconductor work function, we use Equation 3.13 to get E_{iF} and then add the energies from the Fermi level to the vacuum level:

$$E_{iF} = 0.0259 \text{ eV} \ln \frac{10^{17} \text{ cm}^{-3}}{10^{10} \text{ cm}^{-3}} = 0.417 \text{ eV}$$

$$\Phi_B = \frac{1}{q}\left(\chi + \frac{E_g}{2} + E_{Fi}\right) = 4.05 \text{ V} + 0.56 \text{ V} + 0.417 \text{ V}$$

$$= 5.027 \text{ V}$$

The energy band diagram can now be drawn with the materials separated, as shown in Figure 5.12a, using a common vacuum level. For $V_{GS} = 0$ V, the Fermi level in the metal and semiconductor needs to be aligned. This requires moving the metal band diagram down. The vacuum level must be continuous, so it has a positive slope through the insulator and the semiconductor. The complete band diagram is shown in Figure 5.12b. Notice that the NMOS is in weak inversion even though there is zero voltage applied to the gate.

To make the semiconductor energy bands flat will require applying a voltage to make the band diagram look similar to Figure 5.12a. This requires applying a voltage to the gate. Since a positive voltage moves the Fermi level down, a negative voltage is required to move the Fermi level up. The gate voltage required is equal to the difference in the work function between the gate and the semiconductor. That is, applying $V_{GS} = \Phi_G - \Phi_B = \Phi_{GB} = -0.927\ V$ will make the semiconductor energy band flat. This is shown in Figure 5.13.

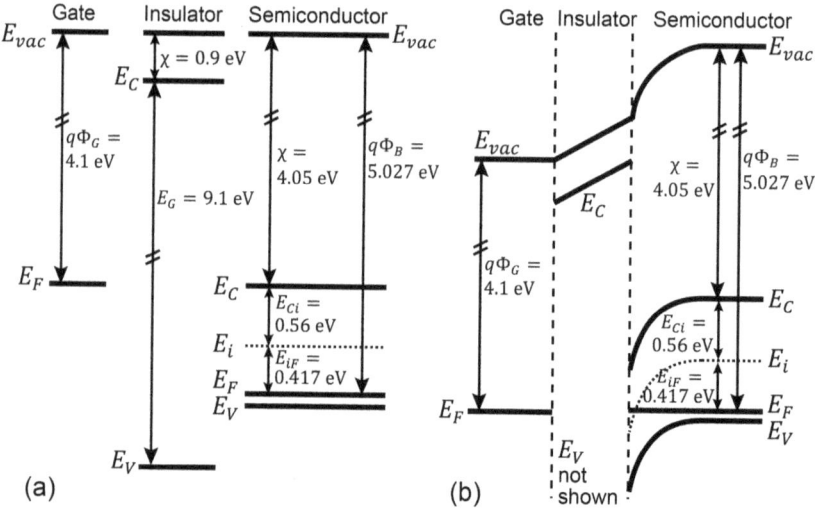

Figure 5.12. Band diagrams for Example 5.1. (a) A band diagram in which the materials are separated. (b) A band diagram with zero gate voltage. This NMOS is in weak inversion with zero voltage applied.

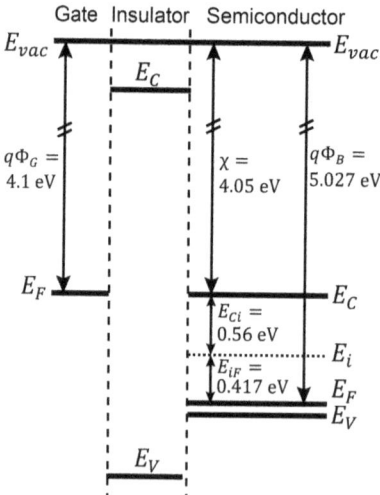

Figure 5.13. Band diagram for Example 5.1 for the flat band condition: $V_{GS} = V_{FB} = -0.927\ V$.

The example above showed us that the voltage required to obtain a flat band diagram, known as the flat band voltage V_{FB}, is:

$$V_{FB} = \Phi_G - \Phi_B = \Phi_{GB} \quad \text{(incomplete equation!)}$$

This equation is not complete. This shows the effect of the work function on the flat band voltage. In the next section another term will be introduced due to the charges within the gate insulator.

5.6.2 Charges in the Gate Insulator

In practice, there are electrical charges within the gate insulator. These charges may be categorized in three categories, as shown in Figure 5.14.

Trapped charges. There are often charges trapped within an insulator since the insulator does not permit the charges to flow to a conductor. These charges could be negative, such as from a stray electron, or they could be positive, due to an electron missing from the valence band (a hole). These trapped charges arise during fabrication as well as during device operation.

Mobile charges. There are often charges that can move. For example, sodium is a very small atom that easily ionized and moves slowly in many insulators, such as SiO_2. The higher the temperature, the faster the sodium can move. One of the reasons semiconductor companies use a cleanroom with ultrapure chemicals is to avoid sodium and potassium contamination. The mobile charges are normally positively charged. With care during fabrication, the mobile charge density can be zero, and therefore we can ignore them in this book.

Surface State charges. All semiconductors have dangling chemical bonds on their surface which are positively charged. The density of charges depends upon how the semiconductor surface is passivated. For example, depositing metal on the semiconductor, depositing an insulator on the semiconductor, and cleaving the semiconductor in a high vacuum will all result in different charge densities for the same semiconductor material. Surface state charges have been studied in detail over the years, and there are many details that will be ignored in this book. For silicon MOSFETs, the surface state charges are the dominant type of charge.

An easy way to understand the effect of these charges on the threshold voltage (V_T) is to start by considering a semiconductor with the appropriate band diagram. The voltage to obtain this diagram was from the charges in the semiconductor. Using the relation, $Q = C \, \Delta V$, we can

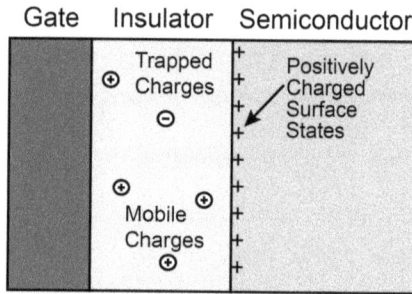

Figure 5.14: Schematic showing the trapped charges, mobile charges, and surface state charges that exist within a semiconductor.

see the effect of the charges. Consider a sheet of charges at the semiconductor interface, and a sheet of charges at the midpoint of the insulating layer.

Charge location:	Middle of insulator	Insulator-semiconductor interface
Distance from gate:	$t_i/2$	t_i
Surface charge density	Q	Q
Gate electrode charge density	$-Q$	$-Q$
Capacitance, $C = \epsilon_i\epsilon_0/d$	$\dfrac{2\epsilon_i\epsilon_0}{t_i}$	$\dfrac{\epsilon_i\epsilon_0}{t_i}$
Change in gate voltage, $\Delta V = \dfrac{-Q}{C}$	$-\dfrac{1}{2}\dfrac{Qt_i}{\epsilon_i\epsilon_0}$	$-\dfrac{Qt_i}{\epsilon_i\epsilon_0}$

This table shows that the effect of the charges is greatest (most negative) when the charges are located the furthest from the gate electrode. If contaminants are controlled during fabrication, the surface state charge density typically dominates over the mobile charge density and trapped charge density. The surface state charge density also has the greatest effect since it is furthest from the gate. Thus, for simplicity, we will assume all the charges are located at the semiconductor surface. To include trapped charges, they can be mathematically mapped to have an equivalent charge density at the semiconductor surface. The density of surface states is represented by N_{SS} and has units of # states/cm^2. The surface state charge density is represented by Q_{SS} and has units of C/cm^2.

Silicon is the most commonly used semiconductor for fabricating MOSFETs. The reason is that silicon has a very low surface state charge density when a layer of SiO_2 grown on it. However, no one has figured out how to get Q_{SS} down to zero. The surface state charge density in silicon is due to chemical bonds that are incomplete, and **the surface state charge is always positive**. Each surface state contributes the charge magnitude due to one electron (1.6×10^{-19} C).

$$Q_{SS} = qN_{SS} \tag{5.32}$$

When making MOSFETs, a value for N_{SS} of 10^{10} charges/cm^2 is considered good and a value of 10^{11} charges/cm^2 is considered bad. Let us see how the surface state charges affect the threshold voltage.

Example 5.2: Consider a MOSFET with a 100 nm thick insulator made of SiO_2. What is the shift in threshold voltage due to a surface state density of (a) 10^{10} charges/cm^2 and (b) 10^{11} charges/cm^2?

Surface state density, N_{SS}	10^{10} charges/cm^2	10^{11} charges/cm^2
Q_{SS}	1.6×10^{-9} C/cm^2	1.6×10^{-8} C/cm^2
ϵ_i	3.9	3.9
C_i	34.5 nF/cm^2	34.5 nF/cm^2
ΔV_T	46 mV	460 mV

As can be seen in Example 5.2, the smaller surface state density has a smaller effect on the threshold voltage. In addition to the effect on the threshold voltage, it has been experimentally shown that the surface charge density negatively affects the electron and hole mobility of a MOSFET, making the drain current smaller and causing CMOS circuits to operate slower.

5.6.3 Gate Materials

The gate must be made of a conductor. One common approach is to use a metal for the gate material. If a metal is used, the work function of the gate may be found from Appendix A.

A second approach is to use heavily doped **polycrystalline silicon** for the gate material. The phrase polycrystalline silicon is often shortened to **polysilicon**, or even just **poly**. Normally we prefer to use single crystal silicon because it has fewer defects and a higher mobility than silicon that is polycrystalline. However, the procedure of fabricating MOSFETs makes it very difficult to obtain a single crystal on top of the amorphous gate insulator, and therefore polycrystalline silicon is used for the gate material.

It is desired to have a good conductor for the gate, but silicon is not a good conductor unless it is heavily doped. There are two options: heavily doped n+ polysilicon, or heavily doped p+ polysilicon. The '+' sign is used to indicate that the doping concentration is very high. The doping concentration is generally so large that the Fermi level in n+ polysilicon may be considered to be at the conduction band edge, and the Fermi level in p+ polysilicon may be considered to be at the valence band edge. The work functions for polysilicon are summarized in the following table.

Material	Work Function
N+ polysilicon	$\Phi_G = 4.05\ V$
P+ polysilicon	$\Phi_G = 4.17\ V$

5.6.4 Flat Band Voltage Summary

The flat band voltage is found by summing up the effects due to the difference in the work function between the gate and semiconductor, and due to the charges within the oxide:

$$V_{FB} = \Phi_{GB} - \frac{Q_{ss}}{C_i} \quad \text{(NMOS and PMOS)} \tag{5.33}$$

This equation assumes that the surface state charge is positive, which it always is. The insulator capacitance per unit area is found from:

$$C_i = \frac{\epsilon_i \epsilon_0}{t_i} \qquad (5.34)$$

where ϵ_i is the relative dielectric constant of the gate insulator, $\epsilon_0 = 8.854 \times 10^{-14}$ F/cm, and t_i is the thickness of the gate insulator. The flat band voltage is used in the equation to calculate the threshold voltage.

5.7 Threshold Voltage Adjustment

The threshold voltage is a very important parameter. During the design of a MOSFET, it is a common problem that we have design constraints on the oxide thickness and substrate doping concentration that will not permit us to obtain the desired threshold voltage. We need a clever trick to permit us to "adjust" the threshold voltage to any value we desire.

Let's recall that one term from the equation for V_T is V_{FB}, from which we can find the following term:

$$V_T = \cdots - \frac{Q_{ss}}{C_i} \qquad (5.35)$$

Here is the trick we will use. We will put charges on the semiconductor surface, simulating a change in Q_{ss}, and adjust the threshold voltage as desired. The charges will be introduced by placing dopant atoms on the semiconductor surface, since the dopant atoms will be ionized. If we put acceptor atoms on the semiconductor surface, which are negatively ionized, then the threshold voltage will shift in the positive direction. If we put donor atoms on the semiconductor surface, which are positively ionized, then the threshold voltage will shift in the negative direction. The equation for the change in threshold voltage is:

$$\Delta V_T = -\frac{Q_{TI}}{C_i} \qquad (5.36)$$

where Q_{TI} is the dopant charge per unit area, not per unit volume, and the subscript 'TI' stands for Threshold adjust Implant. The word implant refers to a machine called an ion implanter that is used to introduce this charge at the surface.

In practice, a machine called an **ion implanter** is used to introduce the dopant atoms to the surface. This machine starts with a gas containing the appropriate dopant atom, uses a plasma to break down the gas into its elemental parts, extracts the ionized dopant atoms, and rasters these dopant atoms on the wafer using electrostatics, similar to the way old CRT tubes rastered an electron beam on a phosphor screen to project an image. These machines can introduce very small quantities of dopant atoms, permitting excellent control of the threshold voltage.

Virtually all MOSFETs made today use the ion implanter to adjust the threshold voltage. This is a very common, and usually necessary, technique to optimize MOSFET design.

5.8 Review of Accumulation, Depletion, Weak Inversion, and Strong Inversion

Table 5.1 summarizes some important differences between accumulation, depletion, weak inversion, and strong inversion. Table 5.1 shows the range of gate voltage for which the MOSFET is in accumulation, depletion, weak inversion, and strong inversion. The MOSFET is turned on when it is in strong inversion because the inversion charge connects the source and drain.

The MOSFET is turned off when the MOSFET is in accumulation, depletion, or weak inversion. There is a depletion charge when the MOSFET is biased in depletion or weak inversion, but a depletion charge cannot carry current because it is made up of ionized dopant atoms that cannot move. There is charge in accumulation, but this charge is of the opposite type as the source and drain, and therefore does not connect the source and drain.

Notice that Table 5.1 does not include a zero voltage in the voltage ranges. This is because the flat-band voltage may be positive or negative, and the threshold voltage may be positive or negative. There is nothing in Equations 5.28, 5.29, or 5.33 that requires a positive or negative answer.

There is a common misconception that an NMOS transistor must have $V_T > 0$. This is not true. In fact, **depletion mode NMOS transistor** is the name of an NMOS with $V_T < 0\ V$. The only thing that must be true is that V_T must be more positive than V_{FB}.

Table 5.1: Comparison of different MOSFET modes.

Accumulation	Depletion	Weak Inversion	Strong Inversion
NMOS: $V_{GS} < V_{FB}$	$V_{GS} > V_{FB}$ $V_{GS} < V_T$	$V_{GS} > V_{FB}$ $V_{GS} < V_T$	$V_{GS} > V_T$
PMOS: $V_{GS} > V_{FB}$	$V_{GS} < V_{FB}$ $V_{GS} > V_T$	$V_{GS} < V_{FB}$ $V_{GS} > V_T$	$V_{GS} < V_T$
Majority carrier concentration increases	Majority carrier concentration decreases	Depletion region extends further	Depletion region is at its maximum size
NMOS: more holes at surface PMOS: more electrons at surface	Depletion region forms	E_i crosses E_F	NMOS: electrons at surface form an inversion layer PMOS: holes at surface form an inversion layer
MOSFET is OFF	MOSFET is OFF	MOSFET is OFF	MOSFET is ON

Similarly, there is a common misconception that a PMOS transistor must have $V_T < 0$. This is also not true. A **depletion mode PMOS transistor** is the name of an PMOS with $V_T > 0\ V$. The only thing that must be true is that V_T must be more negative than V_{FB}.

5.9 Example Threshold Voltage Calculations

The next few pages contain a ton of example calculations to find the threshold voltage.

Example 5.3 (NMOS)

Given values:

Gate material	Aluminum
Insulator material	SiO₂
ϵ_i	3.9
Substrate material	Silicon
Substrate doping	$N_A = 10^{17}$ cm^{-3}
Surface state density, N_{SS}	10^{10} cm^{-2}
Insulator thickness, t_i	10 nm

Calculated values:

$\Phi_G = 4.1$ V (Appendix A)

$$E_{iF} = 0.0259 \text{ eV} \ln \frac{10^{17} \text{ cm}^{-3}}{10^{10} \text{ cm}^{-3}} = 0.417 \text{ eV}$$

(Eq. 3.13)

$$\Phi_B = \frac{1}{q}\left(\chi + \frac{E_g}{2} + E_{iF}\right) = 4.05 \text{ V} + 0.56 \text{ V} + 0.417 \text{ V} = 5.027 \, V$$

(It is helpful to draw a band diagram to see the relative energy levels)

$$\Phi_{GB} = \Phi_G - \Phi_B = 4.1 \text{ V} - 5.027 \text{ V} = -0.927 \text{ V}$$

$$C_i = \frac{\epsilon_i \epsilon_0}{t_i} = \frac{(3.9)(8.854 \times 10^{-14} \text{F/cm})}{10 \times 10^{-7} \text{cm}}$$

$$= 345 \text{ nF/cm}^2$$

$$Q_{SS} = (1.6 \times 10^{-19}\text{C})(10^{10} \text{ cm}^{-2})$$

$$= 1.6 \text{ nC/cm}^2$$

$$V_{FB} = \Phi_{GB} - \frac{Q_{SS}}{C_i} = -0.927 \text{ V} - \frac{1.6 \text{ nC/cm}^2}{345 \text{ nF/cm}^2} = -0.932 \text{ V}$$

$$\phi_{SB} = 2\frac{k_B T}{q}\ln\frac{N_A}{n_i} = 2(0.0259 \text{ V})\ln\frac{10^{17} \text{ cm}^{-3}}{10^{10} \text{ cm}^{-3}} = 0.834 \text{ V}$$

$$V_T = -0.932 \text{ V} + 0.834 + \cdots$$

$$+ \frac{\sqrt{2(11.8)(8.854 \times 10^{-14}\text{F/cm})(1.6 \times 10^{-19}\text{C})(10^{17}\text{cm}^{-3})(0.834\text{V})}}{345 \text{ nF/cm}^2}$$

(Eq. 5.25)

$$V_T = -0.932 + 0.834 + 0.484$$

$$V_T = 0.386 \text{ V}$$

Example 5.4 (NMOS)	
Given values:	
Gate material	P+ Polysilicon
Insulator material	SiN_xO_{1-x}
ϵ_i	4.1
Substrate material	Silicon
Substrate doping	$N_A = 10^{16}$ cm^{-3}
Surface state density, N_{SS}	2×10^{10} cm^{-2}
Insulator thickness, t_i	8 nm

Calculated values:

$$\Phi_G = \frac{1}{q}(\chi + E_g) = 4.05 \text{ V} + 1.12 \text{ } V = 5.17 \text{ V}$$

(See section 5.6.3)

$$E_{iF} = 0.0259 \text{ eV} \ln \frac{10^{16} \text{ cm}^{-3}}{10^{10} \text{ cm}^{-3}} = 0.358 \text{ eV}$$

(Eq. 3.13)

$$\Phi_B = \frac{1}{q}\left(\chi + \frac{E_g}{2} + E_{iF}\right) = 4.05 \text{ V} + 0.56 \text{ V} + 0.358 \text{ V}$$

$$= 4.968 \text{ V}$$

(It is helpful to draw a band diagram to see the relative energy levels)

$$\Phi_{GB} = \Phi_G - \Phi_B = 5.17 \text{ V} - 4.968 \text{ V} = 0.202 \text{ V}$$

$$C_i = \frac{\epsilon_i \epsilon_0}{t_i} = \frac{(4.1)(8.854 \times 10^{-14} \text{F/cm})}{8 \times 10^{-7} \text{cm}}$$

$$= 454 \text{ nF/cm}^2$$

$$Q_{SS} = (1.6 \times 10^{-19} C)(2 \times 10^{10} \text{ cm}^{-2})$$

$$= 3.2 \text{ nC/cm}^2$$

$$V_{FB} = \Phi_{GB} - \frac{Q_{SS}}{C_i} = 0.202 \text{ V} - \frac{3.2 \text{ nC/cm}^2}{454 \text{ nF/cm}^2} = 0.195 \text{ V}$$

$$\phi_{SB} = 2\frac{k_B T}{q} \ln \frac{N_A}{n_i} = 2(0.0259 \text{ V}) \ln \frac{10^{16} \text{ cm}^{-3}}{10^{10} \text{ cm}^{-3}} = 0.716 \text{ V}$$

$$V_T = 0.195 \text{ V} + 0.716 + \cdots$$

$$+ \frac{\sqrt{2(11.8)(8.854 \times 10^{-14} \text{F/cm})(1.6 \times 10^{-19} C)(10^{16} \text{cm}^{-3})(0.716 V)}}{454 \text{ nF/cm}^2}$$

(Eq. 5.25)

$$V_T = 0.195 + 0.716 + 0.108$$

$$V_T = 1.019 \text{ V}$$

This is an NMOS transistor (substrate is doped p-type).

Example 5.5 (PMOS)	
Given values:	
Gate material	N+ Polysilicon
Insulator material	SiO_2
ϵ_i	3.9
Substrate material	Silicon
Substrate doping	$N_D = 10^{17}$ cm^{-3}
Surface state density, N_{SS}	9×10^9 cm^{-2}
Insulator thickness, t_i	5 nm

Calculated values:

$$\Phi_G = \frac{1}{q}(\chi) = 4.05 \text{ V}$$

(See section 5.6.3)

$$E_{Fi} = 0.0259 \text{ eV} \ln \frac{10^{17} \text{ cm}^{-3}}{10^{10} \text{ cm}^{-3}} = 0.417 \text{ eV}$$

(Eq. 3.12)

$$\Phi_B = \frac{1}{q}\left(\chi + \frac{E_g}{2} + E_{iF}\right) = 4.05 \text{ V} + 0.56 \text{ V} - 0.417 \text{ V}$$

$$= 4.193 \text{ V}$$

(It is helpful to draw a band diagram to see the relative energy levels)

$$\Phi_{GB} = \Phi_G - \Phi_B = 4.05 \text{ V} - 4.193 \text{ V} = -0.143 \text{ V}$$

$$C_i = \frac{\epsilon_i \epsilon_0}{t_i} = \frac{(3.9)(8.854 \times 10^{-14} \text{F/cm})}{5 \times 10^{-7} \text{cm}}$$

$$= 691 \text{ nF/cm}^2$$

$$Q_{SS} = (1.6 \times 10^{-19}\text{C})(0.9 \times 10^{10} \text{ cm}^{-2})$$

$$= 1.44 \text{ nC/cm}^2$$

$$V_{FB} = \Phi_{GB} - \frac{Q_{SS}}{C_i} = -0.143 \text{ V} - \frac{1.44 \text{ nC/cm}^2}{691 \text{ nF/cm}^2} = -0.145 \text{ V}$$

$$\phi_{SB} = -2\frac{k_B T}{q}\ln\frac{N_D}{n_i} = -2(0.0259 \text{ V}) \ln\frac{10^{17} \text{ cm}^{-3}}{10^{10} \text{ cm}^{-3}} = -0.834 \text{ V}$$

$$V_T = -0.145 \text{ V} - 0.834 - \cdots$$

$$-\frac{\sqrt{2(11.8)(8.854 \times 10^{-14}\text{F/cm})(1.6 \times 10^{-19}\text{C})(10^{17}\text{cm}^{-3})(0.834\text{V})}}{691 \text{ nF/cm}^2}$$

(Eq. 5.26)

$$V_T = -0.145 - 0.834 - 0.242$$

$$V_T = -1.221 \text{ V}$$

This V_T is negative. This V_T is typical for a PMOS transistor.

Example 5.6: In Example 5.3, the threshold voltage for an NMOS transistor was found to be $V_T = 0.386$ V and $C_i = 345$ nF/cm². Let's say we wanted the threshold voltage to be 0.5 V, but we don't want to change the insulator thickness or the substrate doping concentration. What parameters should we use for a threshold adjustment implant?

Answer: $\Delta V_T = 0.5$ V $- 0.386$ V $= 0.114$ V. This requires charge to make this change is:

$$Q_{TI} = -\Delta V_T C_i = -(0.114 \text{ V})(345 \text{ nF/cm}^2) = -39 \text{ nC/cm}^2$$

Since a negative charge is required, a negative ion, such as Boron, is required. The quantity of Boron atoms required is:

$$N_{TI} = \frac{Q_{TI}}{q} = \frac{39 \text{ nC/cm}^2}{1.6 \times 10^{-19} \text{ C}} = 2.46 \times 10^{11} \text{ cm}^{-2}$$

The answer is that we require a dose of 2.46×10^{11} cm⁻² Boron atoms to be implanted into the channel surface.

5.10 Drain Current

The gate and channel region of a MOSFET is shown in Figure 5.15. On the left is the source and on the right is the drain. There are two voltages being applied: V_{GS} and V_{DS}. We want to calculate the current flowing through the MOSFET, from the drain to the source, but it appears to be a 2-D problem which is difficult to solve. To get an exact answer, the current in a MOSFET has to be solved numerically. We are not doing that in this book.

In order to solve the drain current using a hand analysis, we need to make several approximations. To avoid mathematics in 2D, we will solve the electric potential in a MOSFET using two 1-D problems: the gate voltage will give us an electric potential by solving in the y-direction, and the drain voltage will give us an electric potential by solving in the x- direction. These two electric potentials will then be combined using superposition. This approach is not physically rigorous, but we get useful answer.

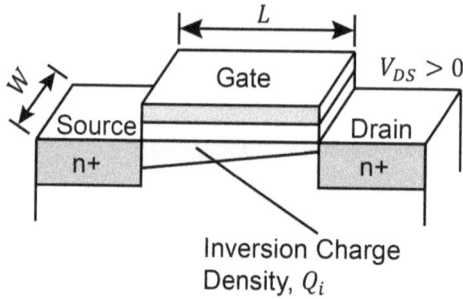

Inversion Charge
Density, Q_i

Figure 5.15:   The channel region of a MOSFET when
operated in strong inversion.   The inversion charge is
shown. The inversion charge is larger on the source side
than the drain side, where the drain voltage reduced the
effective electric potential and therefore reduces the
inversion charge.

Let us examine the NMOS. The inversion charge, made up of electrons,
can carry current. The depletion charge, made up of ionized dopant
atoms, are fixed in place and cannot carry current. The drain current at
any location along the channel may be found from the following
equation:

$$I_{DS} = -qA\mu'_n n(x)\mathcal{E}$$

(5.37)

This equation is based on Equation 3.20. The negative sign arises from
the fact that the current is flowing in the negative x- direction when it
flows from the drain to the source. There are two subscripts on the drain
current to indicate that the current flows to the source, although
technically we only need one subscript to indicate that it is a drain
current. The electron mobility is modified because the electrons are at
the semiconductor surface and therefore experience more scattering
than normal, reducing the mobility from the bulk values reported in
Appendix B. When working with a MOSFET, do not use Appendix B.
All MOSFET problems will have an **effective mobility** given. In the
real world, the effective mobility is measured after a MOSFET is
designed and built. The A in Equation 5.37 is the cross-sectional area
of the inversion charge, which varies with x.

We need to eliminate A. Since the inversion charge Q_i is made up
entirely of electrons, we can relate $n(x)$ to Q_i using $Q_i W = qAn(x)$,

where W is the width of the channel and is known as the **gate width**. Using $\varepsilon = \dfrac{dV}{dx}$, we can rewrite Equation 5.37 as:

$$I_{DS} = -\mu'_n Q_i W \frac{dV}{dx} \tag{5.38}$$

The dx may be moved to the left side of the equation to separate the variables x and V, and then we integrate over the length of the channel:

$$I_{DS} \int_0^L dx = -\mu'_n W \int_0^{V_{DS}} Q_i \, dV \tag{5.39}$$

where L is the **channel length**. The channel length is commonly called the **gate length**, but the channel length is typically slightly smaller than the gate length. For the MOSFETs described in this chapter, the gate length equals the channel length. Carrying out the integration on the left side, this may be simplified to:

$$I_{DS} = -\mu'_n \frac{W}{L} \int_0^{V_{DS}} Q_i \, dV \tag{5.40}$$

The inversion charge may be found by taking the total charge in the semiconductor and subtracting the depletion charge density:

$$Q_i = Q_S - Q_D \tag{5.41}$$

Here we need to make another assumption. We will assume that the depletion charge is a function of V_{GS} only. This is actually a poor assumption, but it allows us to get a simple equation for I_{DS}. If you are interested in a more exact (and much more complicated) equation for the drain current, I recommend the book by Taur and Ning. The semiconductor charge density is found from:

$$Q_S = -C_i \phi_{GS} = -C_i[V_{GS} - V_{FB} - \phi_{SB} - V] \tag{5.42}$$

where the negative sign comes from the fact that we are finding the charge on the negative side of the gate insulator capacitor, ϕ_{GS} represents the voltage across the gate insulator (gate to semiconductor surface), and the voltage across the gate insulator is found using Figure 5.10 as a guide. We had to add the term $(-V)$ because the electric potential is reduced by the drain voltage.

The inversion charge can now be found:

$$Q_i = Q_S - Q_D = -C_i[V_{GS} - V_{FB} - \phi_{SB} - V] - Q_D \tag{5.43}$$

Rearranging terms:

$$Q_i = -C_i\left[V_{GS} - \left(V_{FB} + \phi_{SB} - \frac{Q_D}{C_i}\right) - V\right] \tag{5.44}$$

Using Equation 5.26 for V_T,

$$Q_i = -C_i[V_{GS} - V_T - V] \tag{5.45}$$

Insert Q_i into Equation 5.40 and integrate to obtain the drain current for an NMOS transistor:

$$I_{DS} = \acute{\mu}_n C_i \frac{W}{L}\left[(V_{GS} - V_T)V_{DS} - \frac{1}{2}V_{DS}^2\right] \quad \text{(NMOS)} \tag{5.46}$$

The drain current for a PMOS transistor is calculated in a similar manner, but the voltages are reversed which results in an equation that looks identical except with a negative sign:

$$I_{DS} = -\acute{\mu}_p C_i \frac{W}{L}\left[(V_{GS} - V_T)V_{DS} - \frac{1}{2}V_{DS}^2\right] \quad \text{(PMOS)} \tag{5.47}$$

The mathematics is nice, but let's spend some time understanding what these equations mean. First, keep in mind that we assumed that the MOSFET was turned on. That is, the relation $V_{GS} \geq V_T$ for an NMOS and $V_{GS} \leq V_T$ for a PMOS must be satisfied to use Equations 5.46 or 5.47. When Equations 5.46 or 5.47 are valid, we say that the MOSFET is in the **triode mode**.

Now let's plot Equation 5.46 versus V_{DS} for some gate voltage that turns the NMOS on. This is shown in Figure 5.16a. The curve is a nice parabola, as expected by the V_{DS} and V_{DS}^2 terms. The current goes up, and then the current goes down. Eventually, even with a positive drain voltage, the current becomes negative. This makes no sense. Something must be wrong.

Let us look more closely at Equation 5.45. The maximum value for V is at the drain side of the channel, where $V = V_{DS}$. The inversion charge

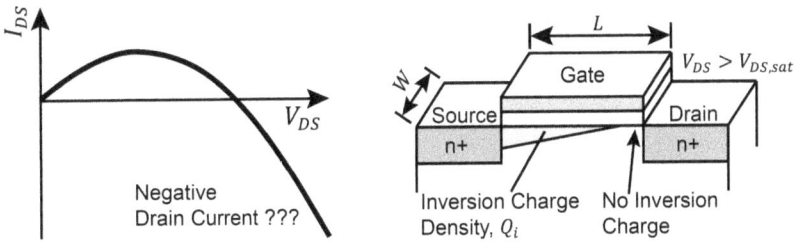

Figure 5.16: (a) Plot of drain current using Equation 5.37 without considering saturation effects. **This plot is wrong!** (b) A large drain voltage causes the inversion layer to disappear at the drain side of the channel (pinched off).

density, Q_i, will become zero if $V_{DS} = V_{GS} - V_T$. Let us give this particular drain voltage a name: $V_{DS,sat}$. Therefore,

$$V_{DS,sat} = V_{GS} - V_T \tag{5.48}$$

When $V_{DS} = V_{DS,sat}$, two things happen: (1) the inversion charge goes to zero at the drain side of the channel and is said to be **pinched off**, and (2) Equations 5.46 and 5.47 reach their maximum value, which may be verified by differentiating the equation and finding where the slope goes to zero. What does it mean when the inversion charge goes to zero? It really means that our assumption that we can solve the 2-D problem using two 1-D problems does not hold. It means that we must solve Poisson's equation in 2-D while simultaneously solving for the charge distribution, which is extremely difficult. What can we do?

Look at Figure 5.16b. The pinched off region is typically very short and the channel at the edge of the pinched off region has a voltage of $V_{DS,sat}$. Therefore, when the inversion charge is pinched off, use Equation 5.46 to find the drain current using $V_{DS} = V_{DS,sat}$. The drain current is independent of V_{DS} when $V_{DS} \geq V_{DS,sat}$. That is, the drain current is a constant for $V_{DS} \geq V_{DS,sat}$. For a PMOS, the drain current is constant for $V_{DS} \leq V_{DS,sat}$. The MOSFET is said to be in **saturation**.

A very important point: Current flows from the drain to the source even when the channel is pinched off. Our simple 1-D equations can't properly describe it, but the current continues to flow.

Table 5.2: MOSFET Drain Current Equation Summary

<table>
<tr><td colspan="2" align="center">Cut-off mode</td></tr>
<tr><td>

$V_{GS} \leq V_T$ (NMOS)

$V_{GS} \geq V_T$ (PMOS)

$$I_{DS} = 0$$
</td></tr>
<tr><td align="center">Triode mode</td></tr>
<tr><td>

$V_{GS} \geq V_T$ and $V_{DS} \leq V_{DS,sat}$ (NMOS)

$V_{GS} \leq V_T$ and $V_{DS} \geq V_{DS,sat}$ (PMOS)

$$I_{DS} = \mu_n C_i \frac{W}{L}\left[(V_{GS} - V_T)V_{DS} - \frac{1}{2}V_{DS}^2\right] \quad \text{(NMOS)}$$

$$I_{DS} = -\mu_p C_i \frac{W}{L}\left[(V_{GS} - V_T)V_{DS} - \frac{1}{2}V_{DS}^2\right] \quad \text{(PMOS)}$$
</td></tr>
<tr><td align="center">Saturation mode</td></tr>
<tr><td>

$V_{GS} \geq V_T$ and $V_{DS} \geq V_{DS,sat}$ (NMOS)

$V_{GS} \leq V_T$ and $V_{DS} \leq V_{DS,sat}$ (PMOS)

$$I_{DS} = \frac{1}{2}\mu_n C_i \frac{W}{L}(V_{GS} - V_T)^2 \quad \text{(NMOS)}$$

$$I_{DS} = -\frac{1}{2}\mu_p C_i \frac{W}{L}(V_{GS} - V_T)^2 \quad \text{(PMOS)}$$
</td></tr>
</table>

Depending upon the applied voltage, the MOSFET is said to be in cut-off, triode, or saturation. Cut-off is when the MOSFET is off. Triode is the mode of operation when the MOSFET is on and the drain voltage affects the drain current. Saturation is the mode of operation in which the drain current does not depend upon the drain voltage. These modes, and the drain current equations for each, are summarized in Table 5.2.

With these equations, we can accurately plot the drain current in a MOSFET. Figure 5.17 shows an example of the drain current versus drain voltage for several gate voltages. As the gate voltage becomes larger (more positive for NMOS, more negative for PMOS), the drain

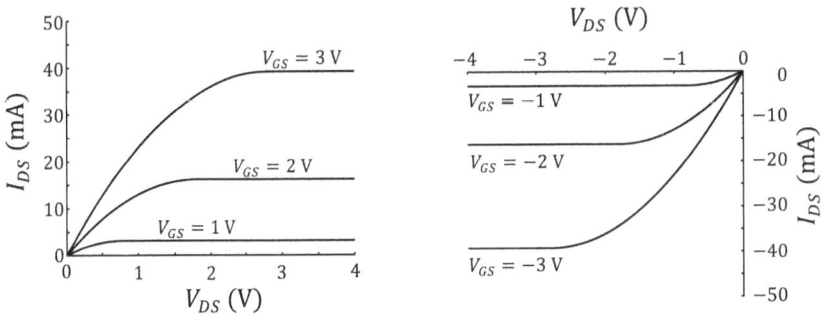

Figure 5.17: Typical family of curves showing the drain current versus drain voltage for a variety of gate voltages. (left) NMOS transistor (right) PMOS transistor.

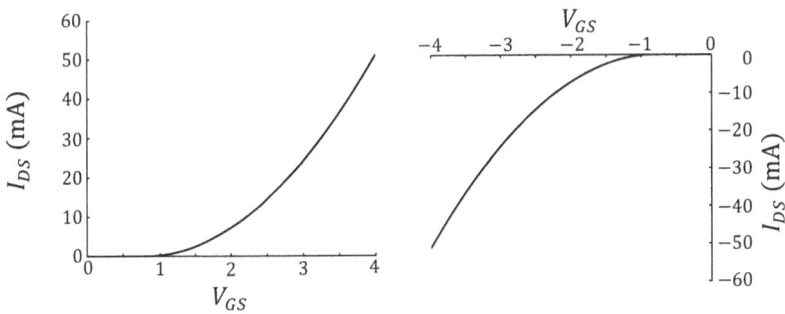

Figure 5.18: Graph of the drain current versus gate voltage when the drain voltage is held constant. (a) NMOS with $V_T = 0.8$ V. (b) PMOS with $V_T = -0.8$ V.

current becomes larger. This figure also shows that the current saturates for large drain voltages. A MOSFET operated in the saturation mode makes for a good current source as the current is independent of the drain voltage. In reality, MOSFETs show an increase in drain current in saturation, especially when the MOSFET is physically very small. This will be discussed in detail in Chapter 11.

Figure 5.18 shows an example of the drain current versus the gate voltage. These graphs were generated for transistors with $V_T = \pm 0.8$ V. A large drain voltage is applied, and it is clear that the drain current has a slight curvature to it.

5.11 Gain: Transconductance and Output Conductance

The gain of an amplifier is the differential change in output divided by the differential change in input. In the case of a MOSFET, the output is the drain current, and the input is either the gate voltage or the drain voltage, as represented in Figures 5.17 and 5.18. The gain of the MOSFET therefore has units of Amps per Volt (A/V), and based on these units we refer to the gain using the term conductance. If the gain is the drain current / gate voltage, we use the term **transconductance** and the symbol g_m. If the gain is the drain current / drain voltage, we use the term **output conductance** and the symbol g_0.

In saturation,

$$g_{m,sat} = \left.\frac{\partial I_{DS}}{\partial V_{GS}}\right|_{sat} = \mu_n' C_i \frac{W}{L}(V_{GS} - V_T) \tag{5.49}$$

$$g_{0,sat} = \left.\frac{\partial I_{DS}}{\partial V_{DS}}\right|_{sat} = 0 \tag{5.50}$$

Having zero output conductance is a great thing for current sources. The transconductance is a function of the gate voltage, and may be increased by increasing the gate voltage and reducing the threshold voltage.

In the triode mode,

$$g_m = \left.\frac{\partial I_{DS}}{\partial V_{GS}}\right|_{triode} = \mu_n' C_i \frac{W}{L} V_{DS} \tag{5.51}$$

$$g_0 = \left.\frac{\partial I_{DS}}{\partial V_{DS}}\right|_{triode} = \mu_n' C_i \frac{W}{L}[(V_{GS} - V_T) - V_{DS}] \tag{5.52}$$

In the triode mode, the transconductance depends on the drain voltage, but in practice the transconductance is limited because a large drain voltage will push the MOSFET into the saturation mode. The output conductance is large for a small drain voltage but goes to zero when the MOSFET is put into the saturation mode. In most analog circuits, it is desired to have a low output conductance.

5.12 Switching Speed

In many applications, it is very important for the transistor to be fast. In digital circuits, the MOSFET should turn on and off very fast. In an analog circuit, it is desired to have a very high bandwidth. In circuits consisting of MOSFETs, there is very little inductance and capacitance is the limiting factor.

One way of looking at the switching speed of a MOSFET is to find the time it takes to charge a capacitor. Since $Q = CV$, and $I = Q \cdot t$, the time to charge a logic gate is:

$$t = \frac{C_{tot}V}{I_{DS}} \tag{5.53}$$

where C_{tot} is the total capacitance of a logic gate being switched on or off. The capacitance is a sum of many capacitances, including the gate capacitance (C_iWL), capacitance between wires, capacitance between the gate and drain, the gate and source, the source and substrate, the drain and substrate, etc.

To obtain a fast circuit, it is desired to minimize the capacitance and maximize the drain current. The switching time dependence on voltage is not obvious because the drain current has a voltage squared term, and thus a **larger voltage decreases the switching time**.

Another way of looking at the switching time is from the RC time constant. This is consistent with Equation 5.53, as they both indicate that the capacitance needs to be minimized. Furthermore, the RC time constant makes it clear that the series resistance must be minimized.

5.13 Temperature Effects

The potential between the MOSFET surface and the body, ϕ_{SB}, changes with temperature. Looking at Equation 5.16, there is a temperature term explicitly written at the front of the equation, but n_i varies exponentially with temperature as described in Chapter 3. The n_i term dominates and causes ϕ_{SB} to decrease as the temperature increases. That is, ϕ_{SB} has a negative temperature coefficient for an NMOS. Since the equation for a PMOS has a negative sign in front, ϕ_{SB} has a positive temperature coefficient for a PMOS. Typical values are a few mV/°C. A larger substrate doping concentration reduces the temperature sensitivity.

How does the flat band voltage change with voltage? Let's start by considering the work function. The work function changes with temperature, but the effect of temperature for each material varies. In some materials, the work function increases with temperature, while in other materials, the work function decreases with temperature. A common scenario is a polysilicon gate on a silicon substrate, making the temperature effect on the electron affinity to cancel since they are the same material. In this case, the change in the flat band voltage is due to the change of E_F within the substrate. For an NMOS, E_F moves away from the center of the band gap as the temperature increases, causing V_{FB} to become more positive. For a PMOS, E_F also moves away from the center of the band gap as the temperature increases, but this movement is towards the top of the band gap, and therefore V_{FB} becomes more negative. The temperature sensitivity decreases as the substrate doping concentration increases.

For an NMOS, the threshold voltage decreases as the temperature increases. This can be seen from Equation 5.27 where the first term has a positive temperature coefficient, but the first term is small. The second term has a negative temperature coefficient, which dominates. The temperature effect in the third term is small due to the square root. In the end, the temperature coefficient of the threshold voltage is typically a few mV/°C in magnitude and negative. For a PMOS transistor, the temperature coefficient of the threshold voltage is typically a few mV/°C in magnitude and positive. That is, for both NMOS and PMOS, an increase in temperature makes it easier to turn the transistor on.

The drain current is proportional to the channel mobility. The temperature dependence of the channel mobility behaves similarly to the bulk mobility described in Chapter 3. The channel mobility typically decreases at temperatures at and above room temperature. Therefore, it is commonly stated that "MOSFETs have a negative temperature coefficient". That is, the drain current decreases as the temperature increases. In reality, the situation is more complicated. As described above, the NMOS threshold voltage becomes smaller and this effect increases the drain current. Thus, it is possible for a MOSFET drain current to have a positive or negative temperature coefficient, depending upon the applied voltage.

Most commonly, the MOSFET drain current has a positive temperature coefficient at a low drain voltage, and a negative temperature coefficient at a large drain voltage.

The transconductance of the MOSFET is reduced as the temperature increases. That is, the gain of the MOSFET decreases with an increase in temperature.

Example 5.6

An NMOS transistor has a threshold voltage of 0.5 V at room temperature (20°C). What is the threshold voltage at 100°C if the temperature sensitivity is -2 mV/°C?

Answer:

The new threshold voltage is:

$$V_T = 500 \text{ mV} - \frac{2 \text{ mV}}{°C}(100°C - 20°C) = 340 \text{ mV}$$

This is a change of 160 mV, and this change is within the specified operating range of a lot of electronic devices.

Now imagine what would happen if the threshold voltage started at 100 mV. The threshold voltage would be negative and the transistor would not turn off with 0 V on the gate!

5.14 Optimal MOSFET Design

Two of the primary goals for MOSFET design are to obtain a high drain current and to obtain a high gain. Looking at the equations for the drain current and the equations for the transconductance, it is apparent that it is desired to do the following:

- Increase the mobility.
- Increase the oxide capacitance, C_i. This is accomplished by using insulators with a high dielectric constant and using very thin insulators.

- Maximize the device width, W. This is easily achieved, but we will see later that the price of a transistor is directly proportional to the width.
- Decrease the gate length, L. A smaller gate length reduces the resistance between the drain and source. Reducing the gate length increases the drain current, the transconductance, and the speed of a circuit because the gate capacitance is reduced. Furthermore, the size of the transistor is smaller, making the MOSFET less expensive. This is a WIN, WIN, WIN, and WIN!
- Increase the drain and gate voltage. However, the voltages are often set by the circuit requirements, and the power increases as the voltage increases.
- Decrease the threshold voltage, V_T. Decreasing the threshold voltage increase the current flow, but the noise margins are negatively impacted in digital circuits and in Chapter 11 we will see that the leakage current increases for small V_T.

5.15 MOSFET SPICE Model

SPICE is a commonly used electrical circuit simulator. To simulate a MOSFET, SPICE uses a model of the MOSFET that accurately predicts the current through the MOSFET based on the drain and gate voltages. The model we have used for the MOSFET is called a LEVEL 1 model in SPICE.

The derivation for the drain current made several assumptions about the inversion charge. A more accurate model for the drain current can be developed and used in SPICE. The accurate models are complicated, and often involve an implicit function. These are fine to solve numerically, but are difficult to work with when performing hand analysis. There are many SPICE models for MOSFETs because one may make many decisions when deciding upon various tradeoffs in computational complexity versus accuracy. In this book we will keep to the simple model appropriate for hand calculations.

5.16 Small MOSFETs

In this chapter, we learned about long-channel MOSFETs. When the MOSFET becomes small, such as is used in all modern ICs, there are other effects that affect the MOSFET performance, and engineers have developed new MOSFET geometries. When the MOSFET becomes small, the source and drain interact with each other. This becomes a problem because it is easy to design a small MOSFET that does not turn off. We will need to learn the material in Chapter 7, p-n junctions, before learning more about MOSFETs in Chapter 11.

5.17 Summary of Key Concepts

- A MOSFET contains a source, drain, gate, and substrate electrode.
- The gate sits on top of an insulating layer, so no current flows through the gate electrode.
- No current flows to the body electrode.
- All the current flows from the drain to the source.
- An NMOS is a n-channel MOSFET. That is, it has n+ regions for the source and drain, and a p-type substrate. When it is turned on, an n-type inversion layer forms.
- A PMOS is a p-channel MOSFET. That is, it has p+ regions for the source and drain, and an n-type substrate. When it is turned on, a p-type inversion layer forms.
- CMOS circuits are used to create both analog and digital circuits. CMOS circuits may contain over 1 billion transistors.
- A MOSFET has four modes: accumulation, depletion, weak inversion, and strong inversion.
- In accumulation, the majority carrier type in the channel becomes larger. In an NMOS, holes are attracted to the surface of the p-type substrate. In a PMOS, electrons are attracted to the surface of the n-type substrate.
- In depletion and weak inversion, a depletion region is formed at the semiconductor surface near the gate.
- In strong inversion, the dominant carrier at the surface is opposite of the doping type. In an NMOS, electrons are at the surface of the semiconductor. In a PMOS, holes are at the surface of the semiconductor.
- A MOSFET must be in strong inversion to form the inversion layer and be "on".
- A MOSFET in accumulation, depletion, or weak inversion is "off".

- The threshold voltage is the voltage required to just bring the MOSFET into strong inversion; the voltage required to turn the MOSFET on.
- The flat band voltage is the gate voltage required to make the energy bands flat in the semiconductor.
- It is possible to adjust the threshold voltage of a MOSFET by implanting positive or negative dopant atoms on the channel surface.
- In the triode region, the drain current increases quadratically with drain current until the drain saturation voltage is reached when it saturates.
- The drain current is constant in saturation.
- The drain current increases with gate voltage.
- In an NMOS, the drain voltage is positive and the drain current is positive.
- In a PMOS, the drain voltage is negative and the drain current is negative.
- The output conductance is the slope of the drain current with respect to the drain voltage.
- The transconductance is the slope of the drain current with respect to the gate voltage.
- A MOSFET in saturation has zero output conductance (infinite output impedance).

5.18 Problems

1. Draw a cross-section of a PMOS transistor and an NMOS transistor. Indicate the type of doping within the semiconductor.

2. What do the following four symbols stand for?
    a. E_{vac}
    b. E_C
    c. E_i
    d. E_V

3. Draw a band diagram of a PMOS in the following modes:
   a. Accumulation
   b. Depletion
   c. Weak Inversion
   d. Strong Inversion

4. Draw a band diagram of an NMOS in the following modes:
   a. Accumulation
   b. Depletion
   c. Weak Inversion
   d. Strong Inversion

5. Plot the charge distribution versus distance for a PMOS, going from the gate to the substrate (through the gate insulator). Make a plot for:
   a. Strong inversion
   b. Weak inversion / Depletion
   c. Accumulation

6. Plot the charge distribution versus distance for an NMOS, going from the gate to the substrate (through the gate insulator). Make a plot for:
   a. Strong inversion
   b. Weak inversion / Depletion
   c. Accumulation

7. A MOSFET has a SiO_2 gate insulator that is 10 nm thick. The silicon substrate is doped with $N_A = 10^{16}$ cm^{-3}. The flat-band voltage is 0.5 V. What are the following?
   a. C_i
   b. ϕ_{SB}
   c. W_{Dm}
   d. Q_D
   e. V_T

8. Can the threshold voltage be negative in an NMOS transistor?

9. Can the threshold voltage be positive in a PMOS transistor?

10. In this problem, we will study how the insulator thickness affects several parameters of a MOSFET. Consider a NMOS with $N_A = 10^{16}$ cm^{-3}, $N_{SS} = 3 \times 10^{10}$ cm^{-2}, and an aluminum gate with $\Phi_G = 4.1$ V. The insulator is made from SiO$_2$. Plot the following parameters versus oxide thickness, ranging from 1 nm to 1 μm:

    a. V_{FB} on a semilog plot.
    b. V_T on a semilog plot.
    c. When the insulator thickness becomes very small, why does the slope of the flat-band voltage become near zero?

11. Consider a MOS structure with a p-type silicon substrate doped with $N_A = 10^{16}$ cm^{-3} boron atoms and an SiO$_2$ thickness of 20 nm. $N_{SS} = 10^{10}$ cm^{-2}. The gate is a polysilicon gate that is heavily doped p-type.

    a. What is the flat band voltage?
    b. What is the threshold voltage?

12. For each of the following gate voltages, is the device from problem #11 in accumulation, depletion, or strong inversion?

    a. $V_{GB} = -3$ V
    b. $V_{GB} = -1$ V
    c. $V_{GB} = 0$ V
    d. $V_{GB} = 1$ V
    e. $V_{GB} = 3$ V

13. Consider a MOS structure with an n-type silicon substrate doped with $N_D = 10^{17}$ cm^{-3} Phosphorus atoms and an SiO$_2$ thickness of 10 nm. $N_{SS} = 2 \times 10^{10}$ cm^{-2}. The gate is a polysilicon gate that is heavily doped n-type.

    a. What is the flat band voltage?
    b. What is the threshold voltage?

14. For each of the following gate voltages, is the device from problem #13 in accumulation, depletion, or strong inversion?

    a. $V_{GB} = -3$ V
    b. $V_{GB} = -1$ V
    c. $V_{GB} = 0$ V
    d. $V_{GB} = 1$ V
    e. $V_{GB} = 3$ V

15. Consider a NMOS with $N_A = 10^{17}$ cm^{-3}, $N_{SS} = 2 \times 10^{10}$ cm^{-2}, and an aluminum gate with $\Phi_G{=}4.1$ V. The insulator is made from SiO$_2$ and is 10 nm thick. The mobility is 350 cm^2/Vs, the gate width is 150 nm, and the gate length is 100 nm. Using a computer, plot the drain current versus the drain voltage, for $V_{DS} = 0$ to 5 V. On one graph, make a plot for $V_{GS} = 0$ V, $V_{GS} = 1$ V, and $V_{GS} = 2$ V.

16. Consider the same MOSFET as problem #15. Using a computer, plot the drain current versus the gate voltage, for $V_{GS} = 0$ to 5 V. Use $V_{DS} = 5$ V.

17. For the MOSFET from problem #15, calculate:
    a. The saturation transconductance for $V_{GS} = 5$ V.
    b. The saturation output conductance for $V_{GS} = 5$ V.
    c. The triode transconductance for $V_{DS} = 1$ V.
    d. The triode output conductance for $V_{DS} = 1$ V, $V_{GS} = 5$ V.

18. Consider a PMOS with $N_D = 10^{17}$ cm^{-3}, $N_{SS} = 1 \times 10^{10}$ cm^{-2}, and a polysilicon gate that is heavily doped p-type. The insulator is made from SiO$_2$ and is 8 nm thick. The surface mobility is 200 cm^2/Vs, the gate width is 150 nm, and the gate length is 100 nm. Using a computer, plot the drain current versus the drain voltage, for $V_{DS} = -5$ to 0 V. On one graph, make a plot for $V_{GS} = 0$ V, $V_{GS} = -1$ V, and $V_{GS} = -2$ V.

19. Consider the same MOSFET as problem #18. Using a computer, plot the drain current versus the gate voltage, for $V_{GS} = -5$ to 0 V. Use $V_{DS} = -5$ V.

20. For the MOSFET from problem #18, calculate:
    a. The saturation transconductance for $V_{GS} = -5$ V.
    b. The saturation output conductance for $V_{GS} = -5$ V.
    c. The triode transconductance for $V_{DS} = -1$ V.
    d. The triode output conductance for $V_{DS} = -1$ V, $V_{GS} = -5$ V.

21. For the MOSFETs shown in problems #15 and #18, calculate the threshold voltage as a function of temperature.

   a. Plot the threshold voltage as a function of temperature from T=200 K to 400 K for both the NMOS and PMOS transistors.

   b. What is the V_T temperature coefficient for the NMOS and PMOS transistors?

22. Consider an unknown NMOS transistor. The following transistor curve is measured while the transistor was forced into saturation. What are:

   a. The threshold voltage

   b. The product of $\mu'_n C_i \frac{W}{L}$. Note that this product is often represented by the symbol k on a discrete MOSFET datasheet.

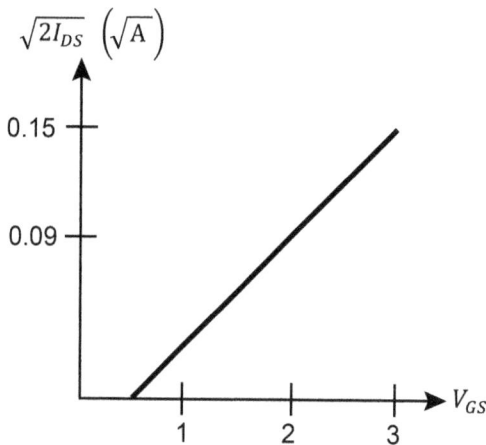

6   Diffusion, Generation, and Recombination

In this chapter, we will learn more about how electrons and holes move and how they interact with each other.

6.1   Why Drift Current Isn't ALL the Current

Early on, we learned about drift current, which obeys Ohm's law. The drift current, assuming the carriers have not reached their saturation velocity, is:

$$J_{\text{drift}} = q(\mu_n n + \mu_p p)\varepsilon \qquad (6.1)$$

In a depletion region, drift current is not sufficient to explain the current that flows. Consider a MOSFET in depletion (no inversion layer). There should be no current between the source and drain because the MOSFET is off ($V_{GS} < V_T$). Pick a point in the depletion region under the gate (any point), and one can easily find that there is an electron concentration, and a hole concentration, and an electric field from the applied voltage. Using Eq. 6.1 we find a non-zero current. And yet, the current must be zero. What is wrong?

Considering only a drift current is not sufficient to explain the current flow in many devices. The equation for the drift current is correct, but there is another type of current, **diffusion current**, which is exactly equal to, but in an opposite direction to, the drift current to produce zero net current in the depletion region of the MOSFET. The diffusion current is an important type of current in many types of semiconductor devices.

6.2   Diffusion Current

The total current is equal to the drift current plus the diffusion current. The definitions of the drift current and diffusion current are:

> **Drift current** is the current that arises due an applied force (the **electric field**) moving electrons or holes.

Diffusion current is the current that arises due to the random movement of the electrons and holes, resulting in a net flow from a **high concentration** to a **low concentration**.

Let us start with an example of diffusion. There are four bins of frogs adjacent to each other: bins A through D. The bins are sitting on a table, and there are 10 frogs in each bin. You hit the table, startling the frogs. Half the frogs jump to the left, and half jump to the right. That is, each bin empties of frogs. Bin A receives 5 frogs from bin B. Similarly, bin B receives 5 frogs from bin A on the left and 5 frogs from bin C on the right. A mathematician looking at the space between any two bins would count the flow of frogs, 5 from the left, and 5 from the right, and find 5 minus 5, or zero frogs jumped between the bins. Ignoring end effects, there is no net flow of frogs when the starting concentration is constant (zero slope). This is illustrated in Figure 6.1.

Now consider the case where bin A has 10 frogs, bin B has 20 frogs, bin C has 30 frogs, and bin D has 40 frogs. This is illustrated in Figure 6.2. You hit the table, and the frogs jump again. Looking between bins A and B, 5 frogs jump to the right and 10 frogs jump to the left, for a net flow of 5 frogs to the left. Looking between bins B and C, 10 frogs jump to the right and 15 frogs jump to the left, for a net flow of 5 frogs to the left. Similarly, between bins C and D, a net of 5 frogs jump to the left.

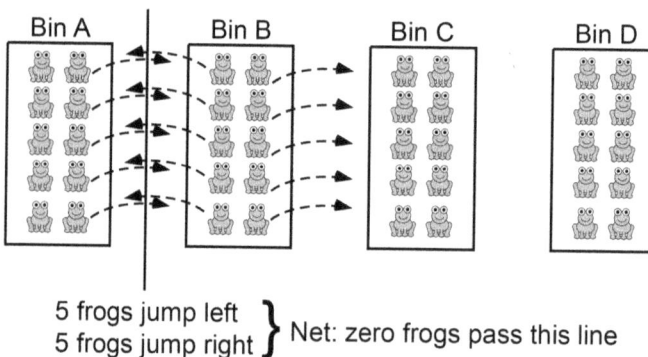

Figure 6.1. Silly example illustrating diffusion. For a constant frog concentration, an equal number of frogs jump left as the number of frogs that jump right, resulting in no net change in frog concentration.

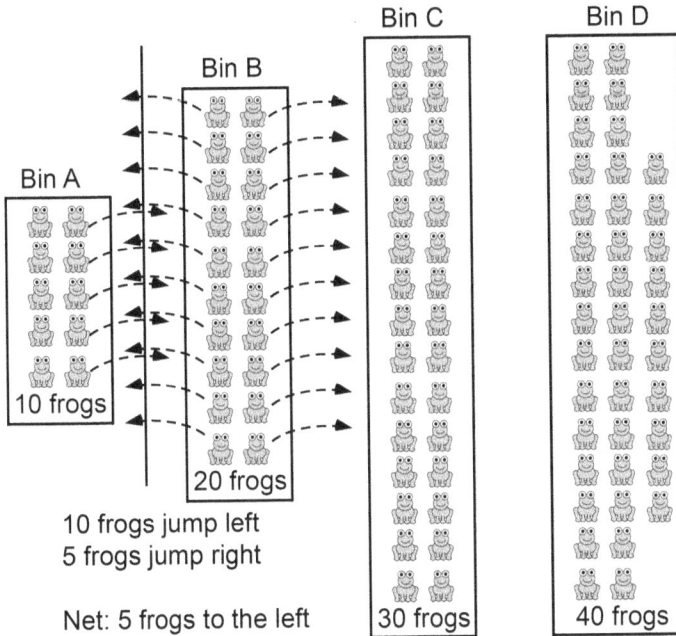

Figure 6.2: With an increasing number of frogs in each bin, the result of the frogs randomly jumping left and right is to cause a net movement of 5 frogs jumping to the left between every bin.

Hence, the net flow of frogs is proportional to the slope of the frog concentration, and not the actual frog concentration.

Electrons and holes are similarly moving in random directions. The spreading of the electrons and holes is called **diffusion**. The net rate of diffusion, or **diffusion flux**, for electrons and holes is:

$$\Theta_n = -D_n \frac{dn}{dx} \tag{6.2}$$

$$\Theta_p = -D_p \frac{dp}{dx} \tag{6.3}$$

where Θ stands for the diffusion flux, D is a diffusion coefficient that relates how rapidly the electrons and holes move, and the subscripts are used to indicate whether it is an electron or a hole that is moving. Notice that the carriers always move from a high concentration to a low concentration. In Figure 6.2, the initial frog concentration increased to the right, indicating a positive slope in the frog concentration, and yet the net motion of the frogs was to the left. Hence, a positive slope results in a negative diffusion flux, necessitating the negative sign in Eqs. 6.2 and 6.3.

The current density due to diffusion, called the **diffusion current density**, is found by multiplying the diffusion flux by the charge of an electron or hole:

$$J_{\text{diff},n} = qD_n \frac{dn}{dx} \qquad (6.4)$$

$$J_{\text{diff},p} = -qD_p \frac{dp}{dx} \qquad (6.5)$$

The different signs arise from the fact that electrons and holes have different charges. This is different from the drift current density, where the electron and hole components have the same sign. There are now four components for the current density:

- electron drift current density,
- electron diffusion current density,
- hole drift current density,
- hole diffusion current density.

$$\boxed{J_{TOT} = J_{\text{drift},p} + J_{\text{drift},n} + J_{\text{diff},p} + J_{\text{diff},n}} \qquad (6.6)$$

The total current density equation and its components can be written in several equivalent ways. We will use whichever version best illustrates the problem being solved at the time.

$$J_{TOT} = q\mu_p p\mathcal{E} + q\mu_n n\mathcal{E} - qD_p \frac{dp}{dx} + qD_n \frac{dn}{dx} \qquad (6.7)$$

$$J_{TOT} = J_{\text{drift}} + J_{\text{diff}} \qquad (6.8)$$

$$J_{\text{drift}} = q(\mu_n n + \mu_p p)\mathcal{E} \qquad (6.9)$$

$$J_{\text{diff}} = qD_n \frac{dn}{dx} - qD_p \frac{dp}{dx} \qquad (6.10)$$

$$J_{TOT} = J_p + J_n \qquad (6.11)$$

$$J_p = q\mu_p p \mathcal{E} - qD_p \frac{dp}{dx} \tag{6.12}$$

$$J_n = q\mu_n n \mathcal{E} + qD_n \frac{dn}{dx} \tag{6.13}$$

The diffusion current density is an effect not seen in conductors. It does not depend on the electric field, and hence it does not obey Ohm's law. That is fantastic, because if the diffusion current density obeyed Ohm's law, we couldn't make devices like diodes that are non-linear with respect to voltage.

Example 6.1: Consider the following graph of electron concentration versus distance. (a) Label the direction of the electron diffusion. (b) Label the direction of electron diffusion current.

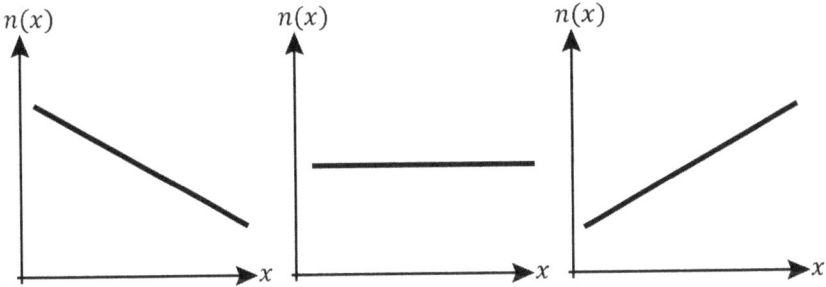

Answer: The direction of diffusion is always from a high concentration to a low concentration. Electrons have a negative charge, and therefore the diffusion current is in the opposite direction of the diffusion direction.

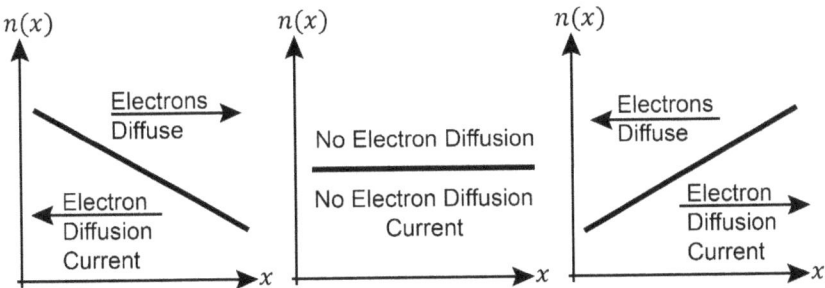

Example 6.2: Consider the following graph of hole concentration versus distance. (a) Label the direction of the hole diffusion. (b) Label the direction of the hole diffusion current.

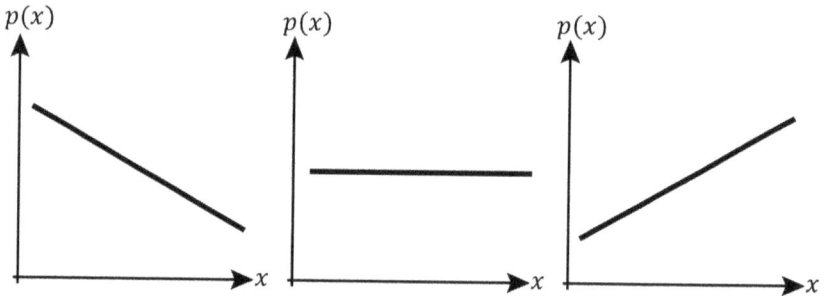

Answer: Holes have a positive charge, and therefore the diffusion current is in the same direction as the hole diffusion.

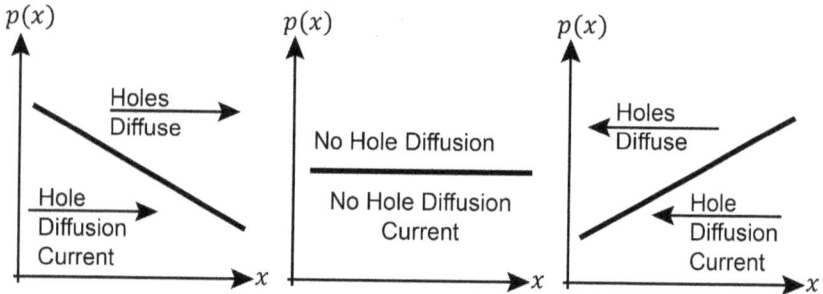

6.3   Einstein Kinetic Relation

There exists a relationship between the carrier mobility and the diffusion constant. This relation was discovered by Albert Einstein and is named after him. It may be found for holes by considering the total hole current density in a semiconductor in equilibrium. An electric field is permitted to exist. The total hole current is zero because we are considering the case of equilibrium. The equation for the hole current is:

$$J_p = 0 = q\mu_p p \mathcal{E} - qD_p \frac{dp}{dx} \tag{6.14}$$

Using the hole concentration at equilibrium:

$$p = N_V \exp\left(\frac{E_V - E_F}{k_B T}\right) \tag{6.15}$$

The slope of the hole concentration is found to be:

$$\begin{aligned}\frac{dp}{dx} &= N_V \exp\left(\frac{E_V - E_F}{k_B T}\right)\frac{d(E_V - E_F)}{dx}\frac{1}{k_B T}\\ &= \frac{p}{k_B T}\frac{d(E_V - E_F)}{dx} = \frac{p}{k_B T}\frac{dE_V}{dx}\end{aligned} \tag{6.16}$$

The dE_F/dx term disappears because the Fermi level must be flat if we are in equilibrium. We can find the electric field from Chapter 4 (see Eq. 4.9):

$$\mathcal{E} = -\frac{d\phi}{dx} = \frac{1}{q}\frac{dE_V}{dx} \tag{6.17}$$

The electric field relates to individual electrons, and therefore we could choose from the three energies E_V, E_i, or E_C, but we could not have used E_F. Now, inserting Equation 6.17 into Equation 6.16, the slope of the hole concentration is:

$$\frac{dp}{dx} = p\mathcal{E}\frac{q}{k_B T} \tag{6.18}$$

Finally, inserting (6.18) back into the current density equation (6.14):

$$J = 0 = q\mu_p p\mathcal{E} - qD_p p\mathcal{E}\frac{q}{kT} \tag{6.19}$$

Canceling terms and rearranging gives:

$$\mu_p = \frac{qD_p}{k_B T} \tag{6.20}$$

A similar result occurs for electrons. Personally, I like to write it as follows, because it rhymes ("D over mu equals kT over Q"):

$$\boxed{\frac{D}{\mu} = \frac{k_B T}{q}} \tag{6.21}$$

The **Einstein kinetic relation** is very useful because it permits us to find the diffusion coefficient directly from the carrier mobility, which we already learned how to calculate in Chapter 3 or Appendix B.

Although we derived Equation 6.21 for a semiconductor in equilibrium, we shall use this result for non-equilibrium cases as well.

Example 6.3: In GaAs, what are the electron diffusion coefficient and the hole diffusion coefficient when the donor concentration is $N_D = 10^{16} \text{ cm}^{-3}$.

Answer: From Table B.4, the electron mobility is $6310 \text{ cm}^2/\text{V} \cdot \text{s}$ and the hole mobility is $367 \text{ cm}^2/\text{V} \cdot \text{s}$. Using Equation 6.21,

$$D_n = \frac{k_B T}{q} \mu_n = 0.0259 \text{ V} \cdot 6310 \, \frac{\text{cm}^2}{\text{V} \cdot \text{s}} = 163.4 \, \frac{\text{cm}^2}{\text{s}}$$

$$D_p = \frac{k_B T}{q} \mu_p = 0.0259 \text{ V} \cdot 367 \, \frac{\text{cm}^2}{\text{V} \cdot \text{s}} = 9.5 \, \frac{\text{cm}^2}{\text{s}}$$

Notice that even though this is an n-type region, there is <u>both</u> an electron and hole mobility, and <u>both</u> an electron and hole diffusion coefficient, because there are both electrons and holes in the n-region.

Example 6.4: Consider an isolated sample of n-type silicon. There are no wires attached, so there can be zero current. The doping concentration on the left is $N_D = 10^{15} \text{ cm}^{-3}$ and the doping concentration on the right is $N_D = 10^{16} \text{ cm}^{-3}$. The doping concentration changes linearly. The sample is 1 mm long. What is the electric field within the silicon due to the diffusion current? Use the average mobility for this region.

Answer: Since the sample is n-type, let us only consider the electrons. First, write an equation for the doping concentration in the form $y = mx + b$:

$$N_D(x) = \frac{10^{16} \text{ cm}^{-3} - 10^{15} \text{ cm}^{-3}}{0.1 \text{ cm} - 0} x + 10^{15} \text{ cm}^{-3}$$

$$N_D(x) = 9.9 \times 10^{16} \text{ cm}^{-4} \, x + 10^{15} \text{ cm}^{-3}$$

Since $n_0 \approx N_D$, we find

$$\frac{dn}{dx} = \frac{dN_D}{dx} = 9.9 \times 10^{16} \text{ cm}^{-4}$$

The average mobility is approximately $1250 \, \frac{\text{cm}^2}{\text{V} \cdot \text{s}}$. The diffusion coefficient is:

$$D_n = \frac{k_B T}{q} \mu_n = 0.0259 \text{ V} \cdot 1250 \frac{\text{cm}^2}{\text{V} \cdot \text{s}} = 32.4 \frac{\text{cm}^2}{\text{s}}$$

The electron diffusion current density is found from Eq. 6.4:

$$J_{\text{diff,n}} = (1.6 \times 10^{-19} \text{ C})\left(32.4 \frac{\text{cm}^2}{\text{s}}\right)(9.9 \times 10^{16} \text{ cm}^{-4})$$

$$= 0.47 \text{ A/cm}^2$$

Since the total current is zero, there must be drift current density with an equal magnitude (and opposite direction) as the diffusion current density.

$$J_{\text{drift,n}} = -0.47 \text{ A/cm}^2$$

From $J_{\text{drift,n}} = q\mu_n n \mathcal{E}$, we get

$$\mathcal{E} = \frac{J_{\text{drift,n}}}{q\mu_n n} = \frac{-0.47 \text{ A/cm}^2}{(1.6 \times 10^{-19} \text{ C})\left(1250 \frac{\text{cm}^2}{\text{V} \cdot \text{s}}\right)(10^{15} \text{ cm}^{-3})}$$

$$= -2.33 \text{ V/cm}$$

on the left side and

$$\mathcal{E} = \frac{J_{\text{drift,n}}}{q\mu_n n} = \frac{-0.47 \text{ A/cm}^2}{(1.6 \times 10^{-19} \text{ C})\left(1250 \frac{\text{cm}^2}{\text{V} \cdot \text{s}}\right)(10^{16} \text{ cm}^{-3})}$$

$$= -0.23 \text{ V/cm}$$

on the right side. The electric field arises to exactly cancel the diffusion current density.

6.4   Concept Review With Zero Current

The last example was quite interesting.   We found that in a semiconductor with zero current flow, that there was an electric field, non-zero drift current density, and non-zero diffusion current density. Let us examine this result conceptually, without the math.

Consider an n-type semiconductor in which the donor doping concentration increases as a function of x.   Since there are more electrons on the right side (positive x), there will be a net diffusion of electrons towards the left side because they move from a high concentration to a low concentration. Since electrons have a negative charge, this results in a diffusion current density towards the right. The electron diffusion and electron diffusion current density are shown on the left side of Figure 6.3. Since the total current is zero, there must be a drift current to the left with the same magnitude as the diffusion current.

Another way to look at this is that there is a net movement of electrons towards the left due to diffusion, which leaves positively charged acceptor atoms on the right. This separation of charge results in an electric field from the right to the left (positive charge to negative charge). The electric field is shown on the right side of Figure 6.3.

A third way of looking at this is to calculate the energy band diagrams due to the doping concentration. On the left side, the semiconductor is lightly doped, and hence the conduction band is further from the Fermi level. On the right side, the semiconductor is heavily doped, and hence the conduction band is close to the Fermi level. The right side of Figure 6.3 shows the energy band diagram. The electrons move from high energy to low energy, and hence move to the right. Since force is:

$$F = -\frac{dE}{dx} \tag{6.22}$$

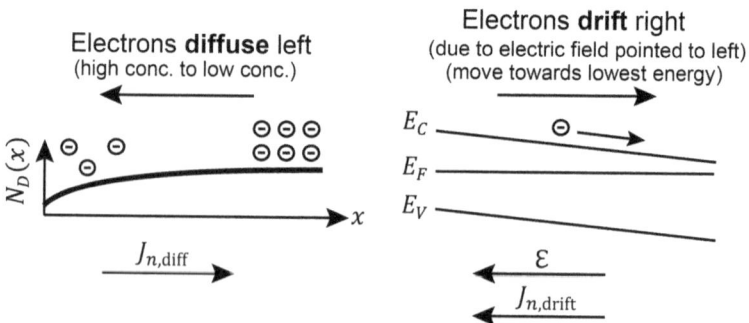

Figure 6.3. Example of n-type semiconductor with an increasing doping concentration. There is a diffusion current and a drift current, but no net current.

the force is directed to the right. This force is the force due to the electric field. Thus, the motion of electrons moving to a lower energy on the right is the drift current.

You can think of the drift current density as being due to the electric field, or the drift current density is being due to the electrons moving from a higher energy to a lower energy. The two descriptions are equivalent.

This example also illustrates the difference between the Fermi level, and the other energy levels, such as E_C. The Fermi level is related to the average energy of the semiconductor, which is constant. Since the Fermi level is constant, there is no net current. The conduction band is related to the energy of individual electrons, which is not constant. Hence, the electric field is related to the slope of E_C. Although there is a force that moves individual electrons towards the right, there is a net flux of electrons towards the left due to diffusion. In the end, the sum of the electrons moving to the right and to the left is zero, resulting in no net current flow and a constant average energy.

The reader may wish to re-read this section several times. Many concepts were combined in a short amount of space. Keep in mind, the different effects are not a consequence of cause and effect. The different effects are all happening simultaneously.

6.5 Recombination and Generation of Carriers

Consider a MOSFET. Under the gate, the electron and hole concentration may change from their equilibrium values, resulting in accumulation, depletion, and inversion. In a diode, we also see the carrier concentrations change under forward or reverse bias. These are examples whereby the carrier concentration can be changed by the application of a voltage. There are other ways of changing the electron and hole concentration. I could shine light on a semiconductor, giving energy to electrons so that they move from the valence band to conduction band. Or I could literally shoot a beam of electrons at the semiconductor, increasing the concentration of energetic electrons in the conduction band. In this section, we will learn how the semiconductor returns to its equilibrium state.

Before discussing electrons and holes, let us discuss how molecules move about in the air we breathe. First, notice that the air molecules are constantly in motion. Second, the energy of individual air molecules vary. Some air molecules are moving fast, and some are moving slowly. Third, the molecules collide with each other, and they collide with objects in a room, such as walls, chairs, books, etc. When a molecule collides, it exchanges energy and momentum with the other particle, and the molecule may increase in energy or decrease in energy. In fact, at atmospheric pressure, the average distance an air molecule travels between collisions is only 60 nm. Individually, these interactions are constantly happening, and the energy of an individual molecule is constantly changing, but the average energy of all the air molecules does not change.

Similar to gas molecules, electrons and holes are also in constant motion, and individually they have varying energies that change with time. In a solid, the electrons and holes are constantly colliding with something. An electron in the conduction band can collide with another electron in the conduction band, or it can collide with a hole in the valence band, or it can collide with a dopant atom, or even collide with an atom that is out of position within the crystal due to thermal vibrations. An electron can interact with a negative charge by repulsion, or it can interact with a position charge through attraction. Either way, it is deflected from its initial trajectory and changes energy. Electrons were used in the description here, but the same arguments apply for holes.

Let us consider some important interactions that may occur. First, as shown in Figure 6.4a, an electron in the valence band may gain energy and jump into the conduction band, leaving a hole in the valence band. We call this **generation** because an electron in the conduction band and a hole were generated (created) where they did not exist before. The second interaction, shown in Figure 6.4b, is where an electron in the conduction band loses energy and occupies an empty quantum state in the valence band. This results in the loss of an electron in the conduction band and the loss of a hole. We call this process **recombination**.

The rate at which electrons and holes recombine, or the **recombination rate**, is designated by the symbol R. It has units of # per volume per time, or cm^{-3}/s. Similarly, the rate at which electrons and holes are generated, or the **generation rate**, is designated by the symbol G with units of cm^{-3}/s.

Generation and recombination are constantly occurring. In thermal equilibrium, the rate at which each process is occurring must exactly equal each other. That is, $G_0 = R_0$. If, say, the generation rate were greater than the recombination rate, then the number of electrons in the conduction band would be constantly increasing, and eventually ALL the electrons would move into the conduction band. Obviously, this cannot happen. Going forward, we will ignore G_0 and R_0 since $G_0 = R_0$, and they cancel out. After all, these processes are always occurring, and the rates do not change.

The generation rate that we will consider is due to an **external** process that moves electrons from the valence band to the conduction band. Since this is an external rate, any equation for the generation rate must describe the external process. An example is the absorption of light. The generation rate depends on both the light intensity and the optical absorption coefficient.

The recombination rate is the rate at which excess electrons and holes recombine. Recombination tends to make the electron concentration and hole concentration approach their equilibrium values. The equation for the **recombination rate** is:

$$R = \frac{n - n_0}{\tau} = \frac{\Delta n}{\tau} \tag{6.23}$$

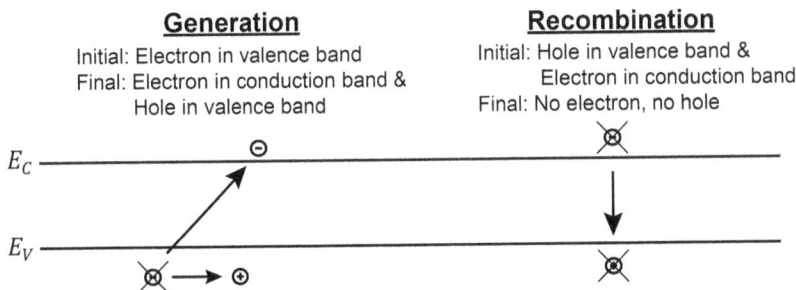

Figure 6.4. Illustration showing how generation and recombination of electrons and holes occurs.

or,

$$R = \frac{p - p_0}{\tau} = \frac{\Delta p}{\tau} \tag{6.24}$$

where τ is the **carrier lifetime**, and represents the characteristic rate at which the electrons and holes recombine. The units for τ are seconds. The carrier lifetime depends upon the number of defects in the semiconductor, as well as details of how the electrons and hole physically recombine. In some semiconductors, such as very pure silicon, the carrier lifetime can be on the order of 1 ms. In other semiconductors, the carrier lifetime may be only 1 ns. Appendix C has some data on carrier lifetimes for silicon, but these numbers are not reliable in the real world. Since the carrier lifetime depends on defects, the carrier lifetime has to be measured.

The values Δn and Δp represent the **excess electron concentration** and the **excess hole concentration**. In equation form:

$$\Delta n = n - n_0 \tag{6.25}$$

$$\Delta p = p - p_0 \tag{6.26}$$

Δn and Δp may be either a positive number or a negative number, because they represent the amount of variation from equilibrium. n and p must both be positive numbers. Thus, the most negative Δn may become is $-n_0$, and the most negative Δp may become is $-p_0$.

In this book we are using a simplistic view of the recombination process whereby (1) the electrons jump directly between the conduction band and valence band and (2) the lifetime is a constant. Reality is much more complicated, but this view is sufficient to describe most semiconductor devices.

Let us consider a few examples:

Example 6.5: A semiconductor is in equilibrium. What are Δn, Δp, and R.

Answer: Since the semiconductor is in equilibrium, $n = n_0$ and $p = p_0$. Then:

$$\Delta n = n - n_0 = n_0 - n_0 = 0$$

$$\Delta p = p - p_0 = p_0 - p_0 = 0$$

$$R = \frac{\Delta n}{\tau} = 0$$

Example 6.6: Light shines on a germanium sample. The light is absorbed at a rate of 10^{16} photons/s/cm^3. Assuming that each photon moves one electron from the valence band to the conduction band, what is the generation rate G?

Answer: The generation rate is $G = 10^{16}$ cm^{-3}s^{-1}.

Example 6.7: A germanium sample has an excess carrier concentration of $\Delta n = \Delta p = 10^{14}$ cm^{-3}. The carrier lifetime is 1 μs. What is the recombination rate?

Answer: The recombination rate is:

$$R = \frac{\Delta n}{\tau} = \frac{10^{14} \text{ cm}^{-3}}{1 \ \mu s} = 10^{20} \text{ cm}^{-3}\text{s}^{-1}$$

6.6 Continuity Equation

We have now seen that carriers move due to an electric field (drift current), carriers move due to diffusion, and electrons and holes move between the conduction band and valence band. We need a set of equations that connects these processes. These equations are called the continuity equations.

The continuity equation is complicated, so let us start with an example that explains the various pieces. We have a large box. On the left and on the right are conveyor belts carrying particles (sand?) in from the left side and out the right side. If the rate of particles coming from the left equals the rate of particles leaving on the right, then there is no change in the quantity of sand in the box. However, if the rate of sand entering from the left is greater than the rate of sand leaving on the right, then sand is accumulating. The rate of sand coming in or leaving will be analogous to electrons or holes entering or leaving the box, which is determined by the current on the left and the current on the right of the box.

In this example, there can also be a hole in the box. Thus, the sand can "disappear" from the box without it being transferred by the conveyor belt. This is analogous to electrons in the conduction band and holes in the valence band "disappearing" due to recombination. Finally, sand can be added to the box from a dump truck. These two events will be called recombination (loss of sand from the box) and generation (addition of sand to the box).

In terms of electrons and holes, the rate of change of the electron or hole concentration may be written as:

$$\frac{dn}{dt} = \frac{1}{q}\frac{dJ_n}{dx} - R + G \tag{6.27}$$

$$\frac{dp}{dt} = -\frac{1}{q}\frac{dJ_p}{dx} - R + G \tag{6.28}$$

These equations are called **continuity equations**. The first term on the right represents the change in electron and hole concentration due to the difference in current (difference in conveyor belt rates). The $(\pm)1/q$ factor is to change the units from charge (Coulombs) to the number of particles. The recombination rate is designated by R (the loss of sand due to the hole at the bottom of the box), and the generation rate is designated by G (the delivery of sand by dump trucks).

This is a deceptively simple equation. If we insert the equation for electron current density into Equation (6.27), we will find that the diffusion term, which already has a derivative, becomes a second derivative. In general, solving the continuity equation involves solving a 2nd order partial differential equation. We will always be looking for methods to simplify the problems through various approximations.

It is common to do an analysis in steady state and with no external generation ($G = 0$). In steady state the time derivatives go to zero. For this case, the continuity equations simplify to:

$$0 = \frac{1}{q}\frac{dJ_n}{dx} - \frac{\Delta n}{\tau} \quad \text{(steady state)} \tag{6.29}$$

$$0 = -\frac{1}{q}\frac{dJ_p}{dx} - \frac{\Delta p}{\tau} \quad \text{(steady state)} \tag{6.30}$$

Example 6.8: Consider a uniformly doped semiconductor in which there is no electric field. Write the hole continuity equation, assuming zero generation and steady state conditions.

Answer: The continuity equation and current density equations are:

$$\frac{d\cancel{p}}{\cancel{dt}} = -\frac{1}{q}\frac{dJ_p}{dx} - R + \cancel{G}$$

$$J_p = \cancel{q\mu_p p \mathcal{E}} - qD_p\frac{dp}{dx}$$

Combining, and substituting $R = \Delta p/\tau$:

$$D_p\frac{d^2p}{dx^2} + \frac{\Delta p}{\tau} = 0$$

But this is in terms of both p and Δp. Note that in a uniformly doped semiconductor,

$$\frac{dp}{dx} = \frac{d(\Delta p - p_0)}{dx} = \frac{d\Delta p}{dx}$$

This allows us to simplify our answer to:

$$\frac{d^2\Delta p}{dx^2} + \frac{\Delta p}{D_p\tau} = 0$$

where we have eliminated p and obtained a relatively simple homogeneous, constant-coefficient 2nd order differential equation. (What is "simple" is always a matter of opinion!)

Example 6.9: Consider a uniformly doped semiconductor in which there is no is no current. The carrier lifetime is 1 μs and the generation rate is $G = 10^{16}$ cm^{-3}s^{-1}. (a) Write the simplified hole continuity equation for steady-state. (b) What is the excess hole concentration?

Answer: (a) The continuity equation in steady state is:

$$\cancel{\frac{dp}{dt}} = -\cancel{\frac{1}{q}\frac{dJ_p}{dx}} - R + G$$

$$0 = -\frac{\Delta p}{\tau} + G$$

(b) Solving for Δp, and inserting numbers, we get:

$$\Delta p = \tau G = 10^{-6}\text{s} \cdot 10^{16}\text{ cm}^{-3}\text{s}^{-1} = 10^{10}\text{ cm}^{-3}$$

Example 6.10: Consider the semiconductor from Example 6.9. The generation source is turned off at time $t = 0$. (a) Find the excess hole concentration for $t > 0$. (b) Plot the excess hole concentration as a function of time.

Answer: (a) We are no longer in steady state, but the current is still zero and the generation rate is zero for $t > 0$. The continuity equation for $t > 0$ is:

$$\frac{dp}{dt} = -\cancel{\frac{1}{q}\frac{dJ_p}{dx}} - R + \cancel{G}$$

$$\frac{d\Delta p}{dt} = -\frac{\Delta p}{\tau}$$

Separating variables and integrating,

$$\int \frac{d\Delta p}{\Delta p} = -\int \frac{dt}{\tau} + \ln C$$

$$\ln(\Delta p) - \ln C = -\frac{t}{\tau}$$

$$\frac{\Delta p}{C} = \exp\left(-\frac{t}{\tau}\right)$$

$$\Delta p(t) = C \exp\left(-\frac{t}{\tau}\right)$$

We used $\ln C$ as the constant of integration. Now we need to solve for C using the boundary condition obtained from the last example: $\Delta p(t = 0) = 10^{10} \text{ cm}^{-3}$.

$$10^{10} \text{ cm}^{-3} = C \exp\left(-\frac{0}{\tau}\right)$$

$$C = 10^{10} \text{ cm}^{-3}$$

The final solution is:

$$\Delta p(t) = 10^{10} \text{ cm}^{-3} \exp\left(-\frac{t}{\tau}\right)$$

(b) The plot is:

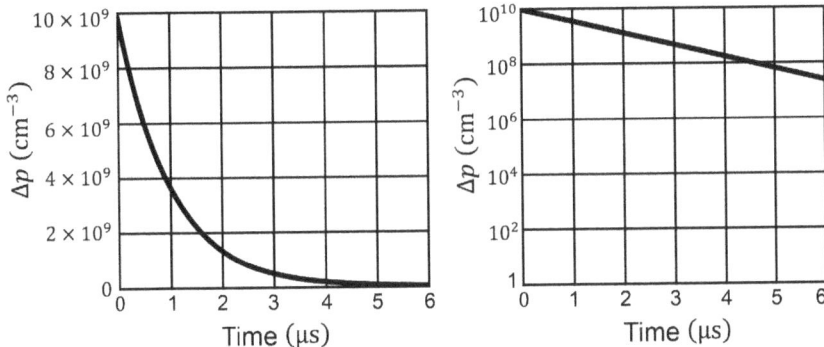

There are two plots here. On the left is a plot on a linear scale, showing that the excess hole concentration drops to near zero after 5 µs. On the right is the same plot, but using a log scale for the y-axis. Now we see that yes, the excess hole concentration has dropped more than 2 orders of magnitude after 6 µs, but it is still rather large. Both plots are correct, and both interpretations are correct! They just highlight different information.

6.7   Diffusion Length

In example 6.8, the parameter $D_p\tau$ cropped up. This is a common parameter that arises when there is diffusion current and the carriers are also recombining. The **electron diffusion length** is defined as:

$$L_n = \sqrt{D_n \tau}$$ (6.31)

And the **hole diffusion length** is defined as:

$$L_p = \sqrt{D_p \tau}$$ (6.32)

The units are length [cm]. The diffusion length may be thought of as the average distance that a carrier (electron or hole) diffuses before recombining.

In the next chapter we will need to use the continuity equation in steady state with only diffusion current. The equation was derived in Example 6.8. Here is the equation again, in terms of the hole diffusion length:

$$\frac{d^2 \Delta p}{dx^2} + \frac{\Delta p}{L_p^2} = 0$$ (6.33)

And for electrons in terms of the electron diffusion length:

$$\frac{d^2 \Delta n}{dx^2} + \frac{\Delta n}{L_n^2} = 0$$ (6.34)

6.8   Summary of Key Concepts

- Drift current is current due to an electric field.
- Diffusion current is due to gradients in the electron or hole concentration.
- The total current is due to the sum of the drift current and the diffusion current.
- The Einstein relation allows us to find the diffusion constants from the carrier mobility.
- An electric field can exist with no applied voltage.
- The recombination rate is the rate at which excess electrons and excess holes recombine.
- The generation rate is the rate at which electrons move from the valence band to the conduction band due to some external stimuli, such as light.

- The excess carrier concentration is the difference between the actual carrier concentration and the equilibrium carrier concentration. For cxample, $\Delta n = n - n_0$.
- Electrons and holes recombine with each other with a characteristic time constant called the carrier lifetime.
- The continuity equation is a fundamental equation that permits one to determine the electron and hole concentration within a semiconductor.
- The continuity equation determines the rate at which the electron or hole concentration is changing within a unit volume of the semiconductor.
- The diffusion length is the average distance that an electron or hole diffuses before recombining.

6.9   Problems

1. Which of the following are examples of drift, and which are examples are diffusion?
   a. Wind is the movement of air molecules from a high pressure to a low pressure.
   b. A person is wearing perfume. A second person sitting nearby can smell the scent of the perfume.
   c. Cream is introduced to a cup of coffee. The cream slowly spreads around, eventually mixing.
   d. Cream is introduced to a cup of coffee. The coffee is stirred to mix the cream and coffee.
   e. A fan moves the air through an air-conditioning duct.
   f. An n-type dopant is applied to the top of a silicon wafer. The wafer is heated to a high temperature, and the dopant atoms move into the silicon.

2. Consider the following plot of electron concentration versus position.
   a. In which direction do the electrons diffuse?
   b. In which direction is the electron diffusion current density?

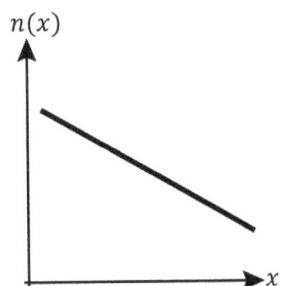

$n(x)$

3. A semiconductor, 10 μm long, has an electron concentration of 10^{15}cm^{-3} on the left, and 10^{17}cm^{-3} on the right. The slope is constant.

    a. What is the electron diffusion current density, assuming $\mu_n = 1000 \ \frac{\text{cm}^2}{\text{V} \cdot \text{s}}$? Use the correct sign in your answer.

    b. Assuming that the net current is zero, what is the electric field on the left? Use the correct sign in your answer.

    c. Assuming that the net current is zero, what is the electric field on the right? Use the correct sign in your answer.

4. A semiconductor, 100 μm long, has a hole concentration of 10^{14}cm^{-3} on the left, and 10^{17}cm^{-3} on the right. The slope is constant.

    a. What is the hole diffusion current density, assuming $\mu_p = 200 \ \frac{\text{cm}^2}{\text{V} \cdot \text{s}}$? Use the correct sign in your answer.

    b. Assuming that the net current is zero, what is the electric field on the left? Use the correct sign in your answer.

    c. Assuming that the net current is zero, what is the electric field on the right? Use the correct sign in your answer.

5. What are (a) the diffusion coefficient for electrons, and (b) the diffusion coefficient of holes, for a GaAs sample doped with $N_A = 10^{17} \text{ cm}^{-3}$.

6. Can the excess electron concentration be negative?

7. Using the continuity equation and diffusion current density equation, find the steady state excess electron concentration for $x > 0$ using the following boundary conditions. You may assume that the electric field is negligible.

$$\Delta n(x = 0) = N$$
$$\Delta n(x = \infty) = 0$$

8. (hard) Do the same problem as #7, but with the following boundary conditions.

$$\Delta n(x = 0) = N$$
$$\Delta n(x = W) = 0$$

Hint: Use $\Delta n(x) = A \cosh\left(\frac{x}{L_n}\right) + B \sinh\left(\frac{x}{L_n}\right)$ for the general solution to the differential equation.

7 Diodes

A diode is made up of a p-type semiconductor connected to an n-type semiconductor. This is a very important device to understand because most semiconductor devices are made using multiple n- and p- type regions. In this chapter, we will learn how p-n junctions operate, and hence understand the operation of a diode. A p-n junction is shown in Figure 7.1 with an electrode at either end. An applied voltage of V_p is applied to the p-region, and an applied voltage of V_n is applied to the n-region. The voltage across the p-n junction is $V_{pn} = V_p - V_n$.

7.1 Energy Band Diagrams

The energy band diagram of a p-n junction can be drawn in the three easy steps shown in Figure 7.2.

For **Step #1**, start with a p-type semiconductor on the left, and an n-type semiconductor on the right. The energy bands do not bend because the materials are separated. The electron affinity $(E_{Vac} - E_C)$ in both regions is the same since there is only one type of material, so the fact that the vacuum level must be continuous also implies that the conduction band must be continuous. Therefore, we can simplify the diagram by removing the vacuum level.

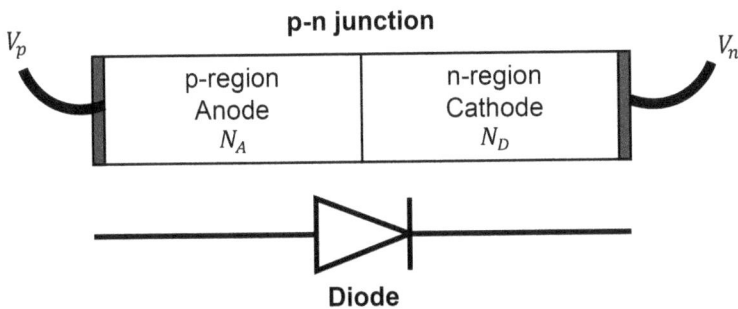

Figure 7.1: A schematic of a p-n junction. A p-n junction is the same things as a diode.

Step #1

Step #2

p-region   n-region

p-region   n-region

Step #3

$x = -x_P$ 0 x_N

p-region   n-region

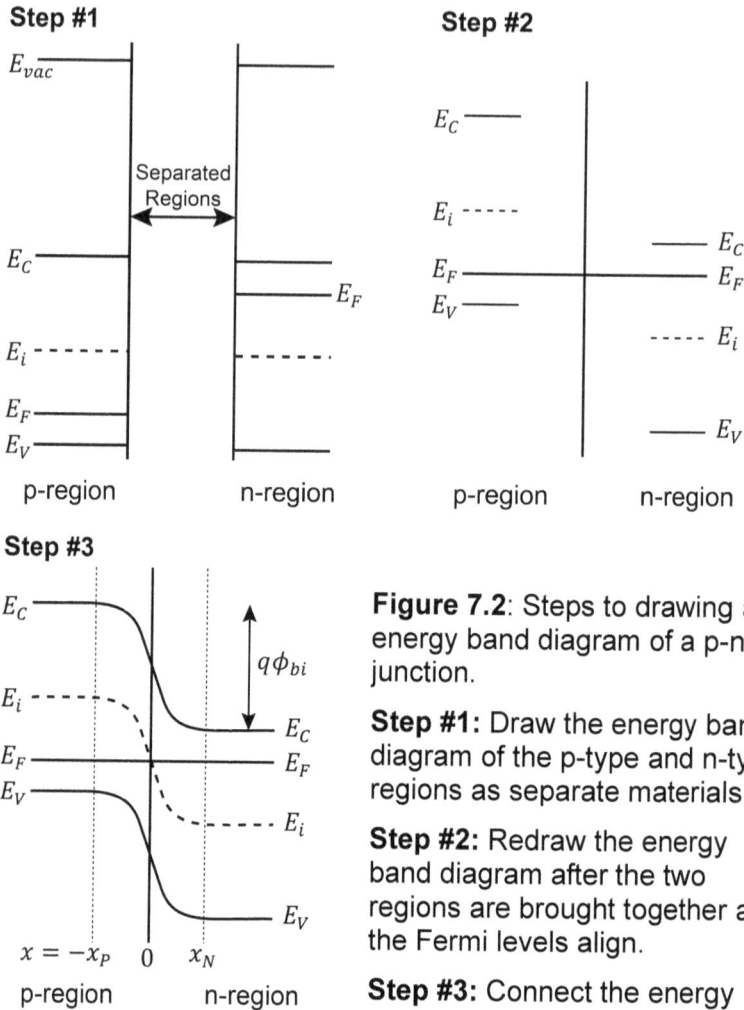

Figure 7.2: Steps to drawing an energy band diagram of a p-n junction.

Step #1: Draw the energy band diagram of the p-type and n-type regions as separate materials.

Step #2: Redraw the energy band diagram after the two regions are brought together and the Fermi levels align.

Step #3: Connect the energy bands near the junction.

For **Step #2**, bring the materials together and align the Fermi levels. Draw the Fermi level across the entire p-n junction as a straight line. Notice that the work function of the p-type semiconductor is greater than that of the n-type semiconductor. Either move the p-region up in energy to align the Fermi levels, or move the n-region down, to align the Fermi levels. These two actions are equivalent because energy is relative. Draw the conduction band and valence band away from the region where the n-region and p-region meet. These are horizontal lines. The energy bands near the junction have not been drawn yet because the energy bands will bend there.

In **Step #3**, the energy bands are drawn near the junction for equilibrium. Keep in mind that both regions are a semiconductor, and therefore the energy bands will bend in both the n-region and the p-region near the junction. The conduction band must be continuous. Later we will prove that the energy bands have a parabolic shape. For the conduction band, start in the p-region and draw a parabolic curve in the p-region, connected to another parabolic curve in the n-region. The curvature is negative in the p-region, and positive in the n-region. The other energy bands are drawn the same way, but the shape is the same so it is easiest just to copy the shape. The difference in electric potential between the p- region and n- region, ϕ_{bi}, is called the built-in potential.

Now consider the case where a voltage is applied between the p- region and the n- region, $V_{pn} = V_p - V_n$. A positive applied voltage $(V_p > V_n)$ is shown in Figure 7.3a, and a negative applied voltage $(V_n > V_p)$ is shown in Figure 7.3b. A positive applied voltage lowers the energy barrier between the p- and n- regions, permitting current to flow. Hence, a positive applied voltage is called **forward bias**. Similarly, a negative applied voltage increases the energy barrier between the p- and n-regions, preventing current from flowing. Hence, a negative applied voltage is called **reverse bias**.

Notice that the energy bands bend in such a way that we always obtain a depletion region, whether we apply a positive voltage or a negative voltage. The p-n junction never goes into accumulation. The origin is located at the interface between the p- and n- regions. The edge of the depletion region in the p- region is located at $x = -x_P$. With this definition x_P is always a positive number. The edge of the depletion region in the n-region is located at $x = x_N$.

Let us revisit the equilibrium case shown in Figure 7.2. An equation for the built-in potential can be determined if we use E_F as a reference energy and find $E_i - E_F$ on either side of the junction:

$$\phi_{bi} = \left.\frac{E_i - E_F}{q}\right|_{P-\text{side}} - \left.\frac{E_i - E_F}{q}\right|_{N-\text{side}} \tag{7.1}$$

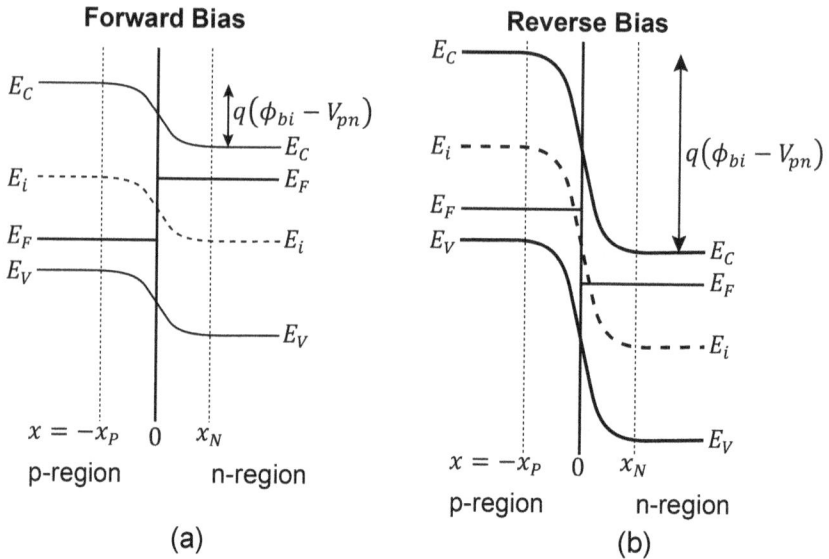

Figure 7.3: (a) A positive applied voltage between the p-region and n- region lowers the electric potential between the two regions. This is forward bias. (b) A negative applied voltage between the p-region and n- region increases the electric potential between the two regions. This is reverse bias.

From Equations 3.12 and 3.13, we find:

$$\phi_{bi} = \frac{k_B T}{q} \ln \frac{N_A}{n_i} + \frac{k_B T}{q} \ln \frac{N_D}{n_i} \tag{7.2}$$

where $p_0 = N_A$ on the p- side and $n_0 = N_D$ on the n- side. Simplifying, the built-in potential is:

$$\boxed{\phi_{bi} = \frac{k_B T}{q} \ln \frac{N_A N_D}{n_i^2}} \tag{7.3}$$

What happens if we apply a large positive voltage? It would appear at first that we could lower the energy bands on the p- region such that the conduction band is lower on the p- region than the n- region, resulting in accumulation instead of depletion. However, when we calculate the current, we will find that the current flow is extremely large when the applied voltage approaches ϕ_{bi}. Practically speaking, we cannot apply a voltage larger than the built-in potential. That is,

$$V_{pn} < \phi_{bi} \qquad \text{(practical limit)}$$

There is no limit on how negative the applied voltage may be, except that eventually the p-n junction will reach breakdown, as will be discussed later.

Example 7.1: What is the built-in potential for (a) a silicon diode, and (b) a GaN diode. Both diodes have $N_A = 10^{15}$ cm^{-3} and $N_D = 10^{18}$ cm^{-3}.

Answer: Using Equation 7.3:

$$\phi_{bi} = 0.0259\,\text{V}\ln\frac{10^{15}\text{ cm}^{-3}\cdot 10^{18}\text{ cm}^{-3}}{(10^{10}\text{ cm}^{-3})^2} = 0.78\text{ V} \quad \text{(a: silicon)}$$

$$\phi_{bi} = 0.0259\,\text{V}\ln\frac{10^{15}\text{ cm}^{-3}\cdot 10^{18}\text{ cm}^{-3}}{(1.77\times 10^{-10}\text{ cm}^{-3})^2} = 3.13\text{ V} \quad \text{(b: GaN)}$$

7.2   Electrostatics

In this section, we will determine things such as the charge, electric potential, and electric field within a p-n junction. We will start with the volume charge density in the depletion region in the n- and p- regions. Then, we will integrate the charge density to obtain the electric field. Then we will integrate again to obtain the electric potential. Finally, using the boundary condition that the electric potential in the n- region is $\phi_{bi} - V_{pn}$, the size of the depletion regions in the n- and p- regions will be found.

The volume charge density is found from:

$$\rho = q[p(x) - n(x) + N_D^+ - N_A^-] \qquad (7.4)$$

The charge density can be broken into three regions: the neutral regions on either side of the depletion, the negative charge in the p-region, and the positive charge in the n-region. Figure 7.4 shows these regions. In the neutral regions, the energy bands are not bending (see Figure 7.2). The Q^- region, located in the p-region, is negatively charged because the energy bands are bending such that the hole concentration is decreasing, leaving the negatively charged N_A^- atoms to dominate the charge in Equation 7.4. Similarly, the Q^+ region, located in the n-

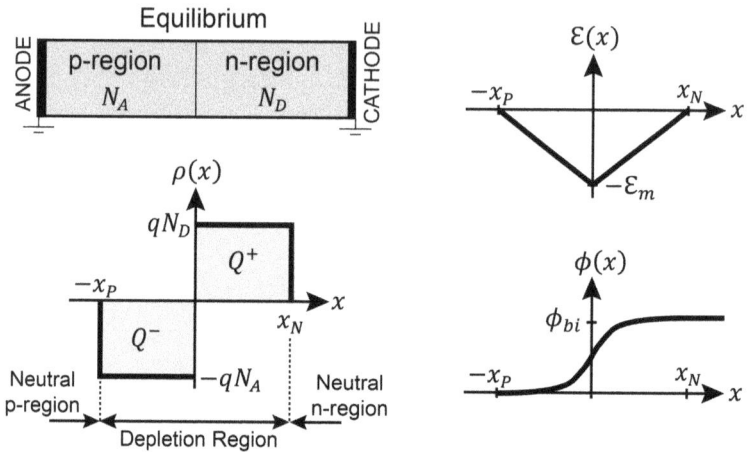

Figure 7.4: Plots for the charge distribution, the electric field, and the electric potential. The depletion region extends into both the p- region and n- region.

region, is positively charged because the energy bands are bending such that the electron concentration is decreasing, leaving the positively charged N_D^+ atoms to dominate the charge in Equation 7.4. The volume charge density is:

$$\rho = \begin{cases} 0 & x \le -x_P, \ x \ge x_N \\ -qN_A^- & -x_P \le x \le 0 \\ qN_D^+ & 0 \le x \le x_N \end{cases} \tag{7.5}$$

The volume charge density is shown in Figure 7.4.

The electric field is found by integrating the charge density using:

$$\mathcal{E}(x) = \frac{1}{\epsilon_s \epsilon_0} \int \rho \, dx \tag{7.6}$$

Since the volume charge density is a constant, the integral is a straight line. In the p-region depletion region, the volume charge density is negative, and hence the electric field is negative. The charge density in the n-region depletion region is positive, and the electric field becomes more positive (less negative), eventually reaching zero at $x = x_N$. This is shown in Figure 7.4.

The electric potential is found by integrating the electric field using:

$$\phi(x) = -\int \mathcal{E}dx \tag{7.7}$$

Since the electric field is a straight line, the electric potential is a quadratic equation. The electric potential may be set to be zero wherever we want. We will set the electric potential to be zero in the neutral p-region. From the energy band diagram, it was found that the electric potential in the neutral n- region reaches ϕ_{bi} in equilibrium, or $\phi_{bi} - V_{pn}$ with an applied voltage, as illustrated in Figure 7.2.

If we do the math, and set the appropriate boundary conditions, the electric field and electric potential are:

$$\mathcal{E}(x) = \begin{cases} 0 & x \leq -x_P, \ x \geq x_N \\ -\dfrac{qN_A^-}{\epsilon_s \epsilon_0}[x + x_P] & -x_P \leq x \leq 0 \\ \dfrac{qN_D^+}{\epsilon_s \epsilon_0}[x - x_N] & 0 \leq x \leq x_N \end{cases} \tag{7.8}$$

$$\phi(x) = \begin{cases} 0 & x \leq -x_P \\ \dfrac{qN_A^-}{\epsilon_s \epsilon_0}\left[\dfrac{1}{2}x^2 + x_P x + \dfrac{1}{2}x_P^2\right] & -x_P \leq x \leq 0 \\ -\dfrac{qN_D^+}{\epsilon_s \epsilon_0}\left[\dfrac{1}{2}x^2 - x_N x + \dfrac{1}{2}x_N^2\right] + \phi_{bi} - V_{pn} & 0 \leq x \leq x_N \\ \phi_{bi} - V_{pn} & x \geq x_N \end{cases} \tag{7.9}$$

At $x = 0$, there are two equations for the electric field and they must equal each other. Similarly, there are two equations for the electric potential at $x = 0$ that must equal each other. These equations are:

$$N_A x_P = N_D x_N \tag{7.10}$$

and

$$\frac{qN_A^-}{\epsilon_s \epsilon_0}\left[\frac{1}{2}x_P^2\right] = -\frac{qN_D^+}{\epsilon_s \epsilon_0}\left[\frac{1}{2}x_N^2\right] + \phi_{bi} - V_{pn} \tag{7.11}$$

These two equations have two unknowns: x_P and x_N. Solving these equations, we obtain:

$$x_P = \sqrt{\frac{2\epsilon_s\epsilon_0}{qN_A}\left(\frac{N_D}{N_D + N_A}\right)(\phi_{bi} - V_{pn})} \qquad (7.12)$$

$$x_N = \sqrt{\frac{2\epsilon_s\epsilon_0}{qN_D}\left(\frac{N_A}{N_D + N_A}\right)(\phi_{bi} - V_{pn})} \qquad (7.13)$$

The total width of the depletion region, $x_P + x_N$, is:

$$W_D = x_P + x_N = \sqrt{\frac{2\epsilon_s\epsilon_0}{q}\left(\frac{N_D + N_A}{N_A N_D}\right)(\phi_{bi} - V_{pn})} \qquad (7.14)$$

The magnitude of the maximum electric field is found by inserting Equation 7.12 into Equation 7.8, setting $x = 0$, and taking the absolute value. This value is designed by $\mathcal{E}_m$.

$$\mathcal{E}_m = \sqrt{\frac{2q}{\epsilon_s\epsilon_0}\left(\frac{N_A N_D}{N_D + N_A}\right)(\phi_{bi} - V_{pn})} \qquad (7.15)$$

If we look carefully at the equations presented here, we can verify the following useful facts:

1. The lightly doped region largely determines the electrostatics of the p-n junction.
2. The depletion width is largest in the lightly doped region.
3. The overall depletion region width is determined by the lightly doped region.
4. The maximum electric field is determined by the lightly doped region.
5. An increase in doping concentration decreases the depletion region width in that region.
6. An increase in the doping concentration of the lightly doped region causes the electric field to increase.
7. An increase in the doping concentration always increases the built-in potential, regardless of whether the doping is on the heavily or lightly doped side.

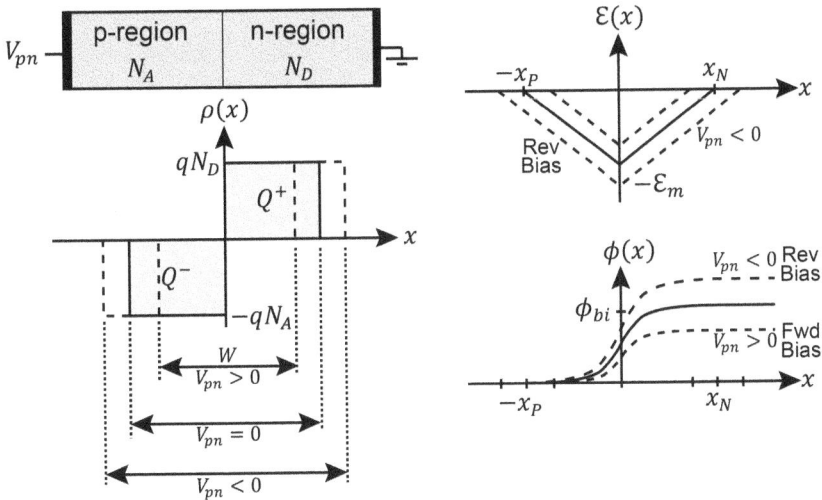

Figure 7.5: Effect of an applied voltage V_{pn} on the depletion region width, total charge Q^+ and Q^-, electric field $\mathcal{E}(x)$, and electric potential $\phi(x)$.

Furthermore, we can verify the following facts related to the applied voltage, which are illustrated graphically on Figure 7.5:

1. Applying a forward bias (positive V_{pn}) causes the depletion region width to decrease, including x_N, x_P, and W.
2. Applying a forward bias (positive V_{pn}) causes the maximum electric field to decrease.
3. Applying a reverse bias (negative V_{pn}) causes the depletion region width to increase, including x_N, x_P, and W.
4. Applying a reverse bias (negative V_{pn}) causes the maximum electric field to increase.

These facts (all 11) are useful concepts as they help us describe how a diode (p-n junction) works without resorting to equations. In future chapters, we will use these concepts a lot. For example, we will see how to adjust the doping concentration to reduce the electric field when we want a high breakdown voltage, or to reduce the size of the depletion region to make very small MOSFETs.

Example 7.2: Consider a silicon diode with $N_A = 10^{15}$ cm^{-3} and $N_D = 10^{18}$ cm^{-3}. How far does the depletion region extend into the n- region and the p- region, what is the total depletion region width, and what is the maximum electric field?

Answer: The built-in potential was found in Example 7.1 to be 0.78 V. Using Equations 7.12 – 7.15:

$$x_P = \sqrt{\frac{2 \cdot 11.8 \cdot 8.854 \times 10^{-14} \text{F/cm}}{1.6 \times 10^{-19} \text{ C} \cdot 10^{15} \text{ cm}^{-3}} \left(\frac{10^{18} \text{ cm}^{-3}}{10^{18} \text{ cm}^{-3} + 10^{15} \text{ cm}^{-3}}\right) 0.78 \text{ V}}$$

$$x_P = 0.0001 \text{ cm} = 1 \text{ μm}$$

$$x_N = \sqrt{\frac{2 \cdot 11.8 \cdot 8.854 \times 10^{-14} \text{F/cm}}{1.6 \times 10^{-19} \text{ C} \cdot 10^{18} \text{ cm}^{-3}} \left(\frac{10^{15} \text{ cm}^{-3}}{10^{15} \text{ cm}^{-3} + 10^{18} \text{ cm}^{-3}}\right) 0.78 \text{ V}}$$

$$x_N = 10^{-7} \text{ cm} = 10^{-3} \text{ μm} = 1 \text{ nm}$$

$$W_D = x_N + x_P \approx 1 \text{ μm} = 1000 \text{ nm}$$

$$\mathcal{E}_m = \sqrt{\frac{2 \cdot 1.6 \times 10^{-19} \text{ C}}{11.8 \cdot 8.854 \times 10^{-14} \text{F/cm}} \left(\frac{10^{18} \text{ cm}^{-3} \cdot 10^{15} \text{ cm}^{-3}}{10^{18} \text{ cm}^{-3} + 10^{15} \text{ cm}^{-3}}\right) 0.78 \text{ V}}$$

$$\mathcal{E}_m = 15,400 \text{ V/cm}$$

This example shows a few things. First, the depletion region, which is approximately 1 μm long, exists mostly on the lightly doped side of the junction. Second, the built-in electric field can be quite high.

Finally, it is imperative to keep track of your units. It will be extremely easy to make a mistake if you don't keep track of units.

7.3 Electron and Hole Concentration Inside the p-n junction

To calculate the current in the p-n junction, we will need to calculate the electron concentration and the hole concentration within the p-n junction. Several approximations will be used to help us:

1. Depletion approximation. The volume charge density is constant within the depletion region and determined by the doping concentration. I.e., $\rho = -qN_A$ or $\rho = qN_D$.
2. Low level injection. The voltages applied will be sufficiently small such that the majority carrier concentration does not

significantly change. Only the minority carrier concentration will change significantly.

3. At the edge of the depletion region in the p-region, $x = -x_P$, there is no electric field and thus no drift current. That is, $\mathcal{E}(x = -x_P) = 0$.

4. At the edge of the depletion region in the n-region, $x = x_N$, there is no electric field and thus no drift current. That is, $\mathcal{E}(x = x_N) = 0$.

First consider the n-region. In the region next to the depletion region, there will be an excess of holes. Assumption #2 says that the majority carrier concentration, the electron concentration, does not change. It is only necessary to calculate the hole distribution. The n-region will be taken to be semi-infinite. That is, the region extends from $x = 0$ to ∞. This is done for mathematical convenience. The n-region, and the excess hole concentration, are shown in Figure 7.4. At $x = \infty$, the excess hole concentration reaches zero. That is, it is far enough from the depletion region that the concentration returns to its equilibrium value: $p = p_0$.

Finding the hole concentration at $x = x_N$ is a little more complicated. First, note that the hole concentration at $x = -x_P$ is very large because it is the majority carrier. That is,

$$p(x = -x_P) = N_A = p_0 \tag{7.16}$$

At $x = x_N$, the hole concentration is very small because it is the minority carrier. With zero voltage applied, the hole concentration is:

$$p(x = x_N) = \frac{n_i^2}{n_0} \tag{7.17}$$

Now divide Eq. 7.17 by Eq. 7.16:

$$\frac{p(x = x_N)}{p(x = -x_P)} = \frac{n_i^2}{n_0(x_N)\, p_0(-x_P)}$$

$$= n_i^2 \frac{1}{n_i \exp\left(\frac{E_F - E_i}{k_B T}\right)\Big|_N} \frac{1}{n_i \exp\left(\frac{E_i - E_F}{k_B T}\right)\Big|_P} \tag{7.18}$$

where we used $n_0 = n_i \exp\left(\frac{E_F - E_i}{k_B T}\right)$ and $p_0 = n_i \exp\left(\frac{E_i - E_F}{k_B T}\right)$. Canceling terms, moving the exponent in the denominator to the numerator, and simplifying, we obtain:

$$\frac{p(x = x_N)}{p(x = -x_P)} = \exp\left(\frac{-(E_i - E_F)|_P + (E_i - E_F)|_N}{k_B T}\right) \tag{7.19}$$

Looking at Equation 7.1, we see that the numerator within the exponent is equal to $-q\phi_{bi}$. Using Equation 7.16, we get:

$$p(x = x_N) = N_A \exp\left(-\frac{q\phi_{bi}}{k_B T}\right) \tag{7.20}$$

For the non-equilibrium case when a voltage is applied, substitute $\phi_{bi} - V_{pn}$ in place of ϕ_{bi}:

$$p(x = x_N) = N_A \exp\left(-\frac{q(\phi_{bi} - V_{pn})}{k_B T}\right)$$
$$= N_A \exp\left(-\frac{q\phi_{bi}}{k_B T}\right)\exp\left(\frac{qV_{pn}}{k_B T}\right) \tag{7.21}$$

Comparing to the equilibrium case, Equation 7.20, the first part of Equation 7.21 is simply the equilibrium hole concentration on the n-side, or n_i^2/N_D. Thus,

$$\boxed{p(x = x_N) = \frac{n_i^2}{N_D} \exp\left(\frac{qV_{pn}}{k_B T}\right)} \tag{7.22}$$

Or,

$$\boxed{\Delta p(x = x_N) = \frac{n_i^2}{N_D}\left[\exp\left(\frac{qV_{pn}}{k_B T}\right) - 1\right]} \tag{7.23}$$

Now let find the excess hole concentration throughout the n-region. We will use the continuity equation in steady state with no drift current. Using Equation 6.33, and repeated here:

$$\frac{d^2\Delta p}{dx^2} - \frac{\Delta p}{L_p^2} = 0 \tag{7.24}$$

This is a standard, constant coefficient 2nd order differential equation. The generic form of the solution may be found in a differential equations book. The generic form of the solution is:

$$\Delta p(x) = C_1 \exp\left(-\frac{x}{L_p}\right) + C_2 \exp\left(\frac{x}{L_p}\right) \qquad (7.25)$$

There are two boundary conditions. The first boundary condition, $\Delta p(x = \infty) = 0$, requires that C_2 must equal zero, or else it is impossible to obtain a value of zero at $x = \infty$. The second boundary condition, Equation 7.23, can then be used to find C_1. After some algebra, the equation for the excess hole concentration is found to be:

$$\boxed{\Delta p(x) = \frac{n_i^2}{N_D}\left[\exp\left(\frac{qV_{pn}}{k_BT}\right) - 1\right]\exp\left(-\frac{x - x_N}{L_p}\right) \qquad x \geq x_N} \qquad (7.26)$$

A similar set of equations may be obtained for Δn:

$$\boxed{\Delta n(x = -x_P) = \frac{n_i^2}{N_A}\left[\exp\left(\frac{qV_{pn}}{k_BT}\right) - 1\right]} \qquad (7.27)$$

$$\boxed{\Delta n(x) = \frac{n_i^2}{N_A}\left[\exp\left(\frac{qV_{pn}}{k_BT}\right) - 1\right]\exp\left(\frac{x + x_P}{L_n}\right) \qquad x \leq x_P} \qquad (7.28)$$

The excess electron concentration and excess hole concentration vary exponentially with distance. In forward bias, they are positive. In reverse bias, they are negative. For an applied voltage more negative than -0.1 V, the voltage exponential term will be negligible compared to "1", and the distribution is independent of voltage.

A plot of the excess electron concentration and excess hole concentration is shown in Figure 7.6 for $N_D > N_A$. Thus, the electron concentration in the p- region is greater than the hole concentration in the n- region. The electron and hole concentration are shown in equilibrium, and the excess electron and hole concentration are shown underneath that graph in equilibrium. In equilibrium, $\Delta n = \Delta p = 0$. On the right are shown the excess electron and hole concentration under forward bias and reverse bias. Also, note that the point with the largest slope, which will give the largest diffusion current, is located in the lightly doped side. This again confirms that the lightly doped side of a p-n junction determines most of the properties of a p-n junction.

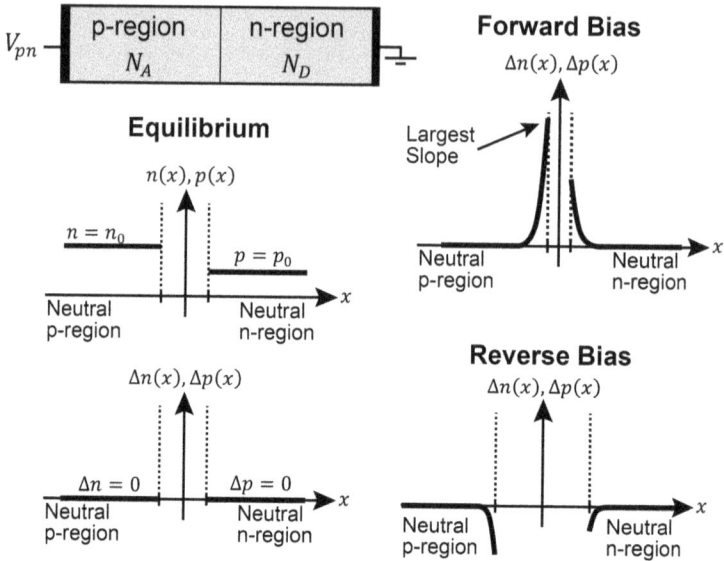

Figure 7.6: The excess electron and hole concentrations are plotted for equilibrium, forward bias, and reverse bias. The absolute electron and hole concentrations are plotted in equilibrium for reference. In this example, $N_D > N_A$. Only the minority carriers are plotted in each graph. In the p-region, only the electron concentration is plotted. In the n-region, only the hole concentration is plotted.

The physical meaning of the **diffusion length**, L_n and L_p, is now apparent. Looking at the exponential decay with distance, the excess carrier concentration drops to 36% of its maximum value at a distance L_n or L_p from the depletion region edge. In other words, the diffusion length represents the decrease of the excess carrier concentration, similar to the time constant of a discharging capacitor as its voltage exponentially decreases.

The **carrier lifetime** has been used, but it is hidden within the diffusion length by $L_n = \sqrt{D_n \tau_n}$ and $L_p = \sqrt{D_p \tau_p}$. Here, I have used τ_n to indicate the carrier lifetime on the p-side, even though it applies to both the electrons and hole on the p-side. This is appropriate because we are finding L_n for the electrons on the p-side. Similarly, τ_p applies to the holes calculated on the n-side.

Example 7.3: Consider a silicon diode with $N_A = 10^{15}$ cm^{-3} and $N_D = 10^{18}$ cm^{-3}. The carrier lifetime in the p- region is 100 μs and the carrier lifetime in the n- region is 10 ns. What are the diffusion lengths of the minority carriers?

Answer: The minority carrier in the p- region are the electrons, and the minority carrier in the n- region are the holes. Thus we will use the acceptor doping concentration when finding the electron mobility, and the donor concentration when finding the hole mobility. Using Table B.3, the mobilities are:

$$\mu_n = 1314 \ \frac{\text{cm}^2}{\text{V} \cdot \text{s}}$$

$$\mu_p = 143 \ \frac{\text{cm}^2}{\text{V} \cdot \text{s}}$$

The diffusion coefficients are found from Equation 6.21:

$$D_n = \frac{k_B T}{q} \mu_n = 0.0259 \ V \cdot 1314 \ \frac{\text{cm}^2}{\text{V} \cdot \text{s}} = 34 \ \frac{\text{cm}^2}{\text{s}}$$

$$D_p = \frac{k_B T}{q} \mu_p = 0.0259 \ V \cdot 143 \ \frac{\text{cm}^2}{\text{V} \cdot \text{s}} = 3.7 \ \frac{\text{cm}^2}{\text{s}}$$

The diffusion lengths are found from Equations 6.31 and 6.32:

$$L_n = \sqrt{D_n \tau_n} = \sqrt{34 \ \frac{\text{cm}^2}{\text{s}} \cdot 10^{-4} \ \text{s}} = 0.058 \ \text{cm} = 580 \ \mu\text{m}$$

$$L_p = \sqrt{D_p \tau_p} = \sqrt{3.7 \ \frac{\text{cm}^2}{\text{s}} \cdot 10^{-8} \ \text{s}} = 1.92 \times 10^{-4} \ \text{cm} = 1.92 \ \mu\text{m}$$

This shows that the diffusion length is much larger on the lightly doped side. The long diffusion length is really determined by the long carrier lifetime. In materials with a very short lifetime, the diffusion lengths are also very small.

Also notice that we calculated parameters for the holes on the n-side, and parameters for the electrons on the p-side. This is because we generally care about the minority carriers when we are working with a p-n junction.

Avoid a common mistake! When finding L_n, we used τ from the p-side, where electrons are the minority carrier! And when finding L_p, we used τ from the n-side where holes are the minority carrier.

Example 7.4: As a continuation of Example 7.3, consider a silicon diode with $N_A = 10^{15}$ cm^{-3} and $N_D = 10^{18}$ cm^{-3}. The carrier lifetime in the p- region is 100 μs and the carrier lifetime in the n- region is 10 ns. What are the excess minority carrier concentrations next to the depletion region for (a) an applied voltage of zero, (b) an applied voltage of 0.5 V, and (c) an applied voltage of -10 V?

Answer: (a) If the applied voltage is zero, the excess carrier concentrations will all be zero. To find the excess minority carrier concentrations next to the depletion region with an applied voltage, Equations 7.24 and 7.28 will be used.

(b) For an applied voltage of 0.5 V, which corresponds to forward bias:

$$\Delta p(x = x_N) = \frac{n_i^2}{N_D}\left[\exp\left(\frac{qV_{pn}}{k_BT}\right) - 1\right]$$

$$\Delta p(x = x_N) = \frac{10^{20}\ cm^{-6}}{10^{18}\ cm^{-3}}\left[\exp\left(\frac{0.5\ V}{0.0259\ V}\right) - 1\right] = 2.42 \times 10^{10}\ cm^{-3}$$

$$\Delta n(x = -x_P) = \frac{n_i^2}{N_A}\left[\exp\left(\frac{qV_{pn}}{k_BT}\right) - 1\right]$$

$$\Delta n(x = -x_P) = \frac{10^{20}\ cm^{-6}}{10^{15}\ cm^{-3}}\left[\exp\left(\frac{0.5\ V}{0.0259\ V}\right) - 1\right] = 2.42 \times 10^{13}\ cm^{-3}$$

For both values, the excess carrier concentration is significantly smaller than the majority carrier concentrations. Thus, the low-level injection approximation is valid. Had a voltage of 0.6 V been used, low-level injection approximation would not be valid! If we encounter situations where the low-level approximation is invalid, we will proceed anyway.

(c) For an applied voltage of -10 V, which is reverse bias:

$$\Delta p(x = x_N) = \frac{10^{20}\ cm^{-6}}{10^{18}\ cm^{-3}}\left[\exp\left(\frac{-10\ V}{0.0259\ V}\right) - 1\right] = -100\ cm^{-3}$$

$$\Delta n(x = -x_P) = \frac{10^{20}\ cm^{-6}}{10^{15}\ cm^{-3}}\left[\exp\left(\frac{-10\ V}{0.0259\ V}\right) - 1\right] = -10^5\ cm^{-3}$$

First, the excess carrier concentrations are always negative in reverse bias. Second, if we calculate p_0 and n_0, we find that $\Delta p = -p_0$ and $\Delta n = -n_0$. Put another way, the electron and hole concentrations go to zero at the edge of the depletion region in reverse bias.

7.4   p-n Junction Current Density

We are now ready to calculate the current in a p-n junction. Because we are clever, we will choose to calculate the current at $x = -x_P$ and $x = x_N$ where the electric field is zero (assumptions #3 and #4 from the last section). Since the electric field is zero here, the current is only diffusion current, and the drift current is zero. This makes the math much easier. We will also use the fact that there is negligible recombination in the depletion region, allowing us to make the claim that the total current through the p-n junction is the sum of the electron diffusion current density in the p-region, and the hole diffusion current density in the n-region. That is,

$$J = J_{\text{diff},n}(x = -x_P) + J_{\text{diff},p}(x = x_N) \tag{7.29}$$

To find the current, we will differentiate $n(x)$ and $p(x)$ to find dn/dx and dp/dx. Then the diffusion equations will be used to find the total current density.

To determine the current density due to holes at the edge of the depletion region in the n-region, differentiate Δp, use $\frac{dp}{dx} = \frac{d\Delta p}{dx}$ to insert it into the diffusion equation (Equation 6.5), and set $x = x_N$:

$$J_p(x = x_N) = q \frac{D_p}{L_p} \frac{n_i^2}{N_D} \left[\exp\left(\frac{qV_{pn}}{k_BT}\right) - 1\right] \tag{7.30}$$

If we do the same procedure for the electrons on the p-side of the p-n junction, we obtain:

$$J_n(x = -x_P) = q \frac{D_n}{L_n} \frac{n_i^2}{N_A} \left[\exp\left(\frac{qV_{pn}}{k_BT}\right) - 1\right] \tag{7.31}$$

The total current density in a p-n junction is the sum of these two currents:

$$J = q n_i^2 \left(\frac{D_p}{L_p N_D} + \frac{D_n}{L_n N_A}\right) \left[\exp\left(\frac{qV_{pn}}{k_BT}\right) - 1\right] \tag{7.32}$$

It is convenient to define a **reverse bias saturation current density**, or simply the **saturation current density**, as:

$$\boxed{J_S = q n_i^2 \left(\frac{D_p}{L_p N_D} + \frac{D_n}{L_n N_A}\right)} \tag{7.33}$$

The current density can be rewritten as:

$$J = J_S \left[\exp\left(\frac{qV_{pn}}{k_B T}\right) - 1 \right] \qquad (7.34)$$

Under reverse bias, the applied voltage is negligible because the current density will become $J = -J_S$. In forward bias, the "-1" may be neglected because the exponent will be much larger. Only for voltages within one or two tenths of a volt does the entire equation need to be used.

Figure 7.7 shows the current density as a function of applied voltage. All four curves in Figure 7.7 show the same current density. They are just scaled to show different aspects of the diode current. The top left shows the current with both forward and reverse bias. The current in reverse bias cannot be discerned because it is very small compared to the forward bias current. This is why it is commonly stated that no current flows in reverse bias.

In reality, current flows in reverse bias, as shown in the top right of Figure 7.7. We can see the current because the forward bias current is not shown. The current density in reverse bias is constant except within a few tenths of a volt from zero. In reverse bias, the current density is $J = -J_S$.

Two more graphs show the current density in forward bias; one using a linear scale and one using a log scale. Under forward bias, the current density is small for small voltages, but increases exponentially, as best seen using the linear scale. However, looking at the log scale for forward bias, there is a substantial current even for a small applied voltage. It is just difficult to see it compared to the extremely large current density shown on the axis used for the linear scale.

The p- n junction makes a good diode. An ideal diode has zero reverse bias current, and the forward bias current shoots up at a very small voltage, called the **turn-on voltage**. The turn-on voltage is really something observed by humans and has no mathematical foundation. Figure 7.8 shows the current versus voltage for a single p-n junction, but for two different y-axis scales. If the full scale range for the current density is 1 A/cm^2, then the turn-on voltage appears to be 0.6 V.

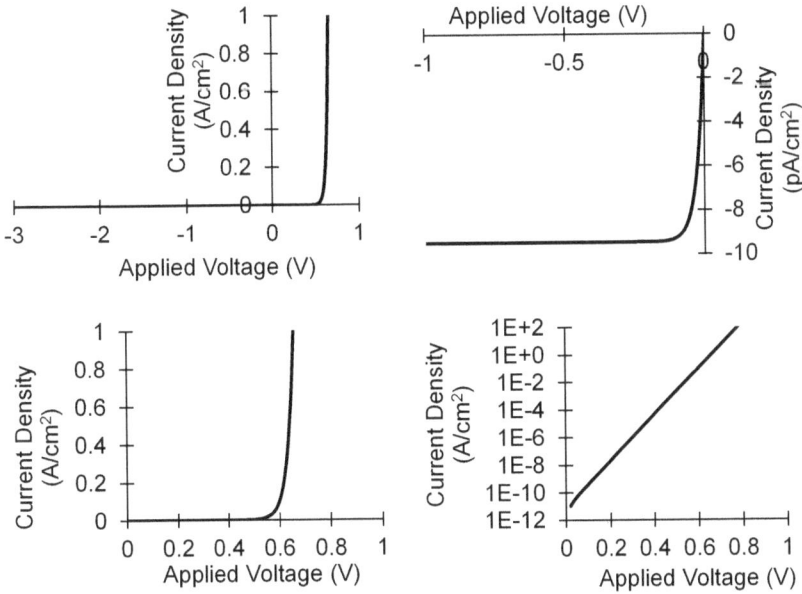

Figure 7.7: All four plots show the same current density for a p-n junction with $J_S = 9.5$ pA/cm^2. The top left shows the current density in both reverse and forward bias. The top right shows only the reverse bias current density. The bottom left shows the current density in forward bias. Notice that it appears that there is very little current until the applied voltage approaches 0.6 V. In reality, as shown in the lower right, where the y-axis has a log scale, there is current for all voltages under forward bias.

However, if the full scale range is 1000 A/cm^2, then the turn-on voltage appears to be 0.8 V. That is the beauty of the exponential function.

For the same size diode with similar doping concentrations, a diode made with a small band gap semiconductor will typically appear to provide a smaller turn-on voltage than for a diode made with a large band gap semiconductor. This is because n_i^2 is orders of magnitude larger for the small band gap semiconductor, as can be verified from the tables in Appendix A. The reverse saturation current depends on n_i^2, and the other parameters will be similar (typically within a factor of 10) of each other, so the variation of n_i^2 dominates. Thus, it is common to

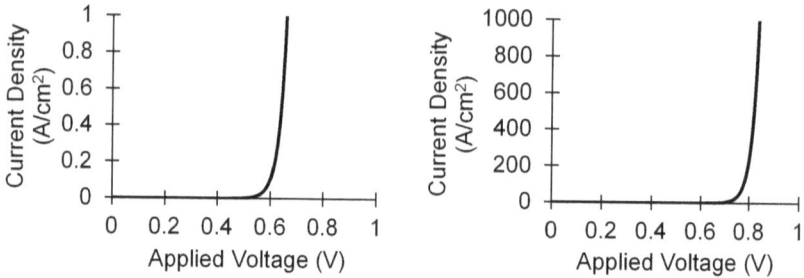

Figure 7.8: These two graphs show the same current versus voltage curves for an identical diode, but the y-axis scale is different. On the left, the turn-on voltage appears to be 0.6 V, while on the right the turn-on voltage appears to be 0.8 V. Since it is the same diode, the turn-on voltage must be a value that depends upon the typical current in the circuit where the diode is used.

hear people say that a Germanium diode has a turn-on voltage around 0.2 or 0.3 V, while a Silicon diode has a turn-on voltage around 0.7 or 0.8 V, and a GaAs diode has a turn-on voltage around 1.4 V.

Example 7.5: As a continuation of Example 7.4, consider a silicon diode with $N_A = 10^{15}$ cm^{-3} and $N_D = 10^{18}$ cm^{-3}. The carrier lifetime in the p- region is 100 μs and the carrier lifetime in the n-region is 10 ns. What are (a) the reverse saturation current density, (b) the current density for an applied voltage of 0.5 V, and (c) the current density for an applied voltage of -10 V?

Answer: The reverse saturation current density is found from Equation 7.33. D_p, D_n, L_p, and L_n were found in Example 7.3.

$$J_s = q n_i^2 \left(\frac{D_p}{L_p N_D} + \frac{D_n}{L_n N_A} \right) = 1.6 \times 10^{-19} \text{ C} \cdot 10^{20} \text{cm}^{-6} \times$$

$$\left(\frac{3.7 \text{ cm}^2/\text{s}}{1.92 \times 10^{-4} \text{ cm} \cdot 10^{18} \text{ cm}^{-3}} + \frac{34 \text{ cm}^2/\text{s}}{0.058 \text{ cm} \cdot 10^{15} \text{ cm}^{-3}} \right)$$

$$J_s = 9.69 \text{ pA/cm}^2$$

Notice that all the variables in the first term in the parenthesis relate to the n- region, even though some variables are subscripted with a 'p' for holes, because p-n junctions are a minority carrier device and holes are a minority carrier in the n-region. Similarly, all the variables in the second term relate to the p- region.

With 0.5 V applied, the current density is:

$$J = J_S \left[\exp\left(\frac{qV_{pn}}{kT}\right) - 1 \right] = 9.69 \text{ pA/cm}^2 \left[\exp\left(\frac{0.5 \text{ V}}{0.0259 \text{ V}}\right) - 1 \right]$$

$$J = 2.35 \text{ mA/cm}^2$$

With -10 V applied, the current density is:

$$J = 9.69 \text{ pA/cm}^2 \left[\exp\left(\frac{-10 \text{ V}}{0.0259 \text{ V}}\right) - 1 \right]$$

$$J = -9.69 \text{ pA/cm}^2$$

In reverse bias, the current density is equal to the negative of the reverse saturation current density. The exponential term above is nearly zero, especially compared to the number "-1".

7.5 Series Resistance

As we analyzed the p-n junction, we focused solely on the junction itself. Current has to flow from the anode electrode, though the p-region, through the junction, through the n- region, and into the cathode electrode. Our analysis never considered the current through the neutral p- region or the neutral n- region. This is acceptable as long as the voltage drop across those regions is negligible. What is negligible? Let's do a quick analysis:

$$\frac{dI}{dV_{pn}} = \frac{qI_S}{k_BT} \exp\left(\frac{qV_{pn}}{k_BT}\right) = \frac{\Delta I}{\Delta V_{pn}} \tag{7.35}$$

The percent change in current is:

$$\frac{\Delta I}{I} \times 100\% = \frac{\frac{qI_S}{k_BT} \exp\left(\frac{qV_{pn}}{k_BT}\right) \Delta V_{pn}}{I_S \left[\exp\left(\frac{qV_{pn}}{k_BT}\right) - 1 \right]} \times 100\% \tag{7.36}$$

The series resistance will be important when a lot of current is flowing. Thus, we can simplify the denominator by eliminating the "-1":

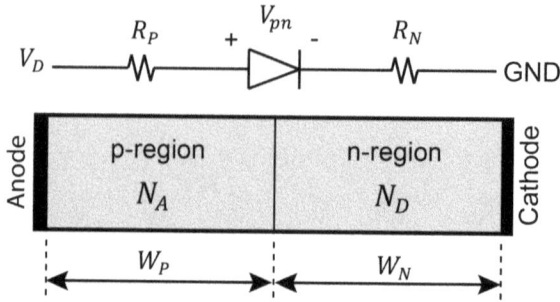

Figure 7.9: Model of a real diode composed of the series combination of two resistors and an ideal diode.

$$\frac{\Delta I}{I} \times 100\% \approx \frac{q}{k_B T} \Delta V_{pn} \times 100\% \tag{7.37}$$

This equation shows that if the applied voltage increases by $\frac{k_B T}{q}$, or 25 mV, then the current will approximately double. Therefore, to have an impact on the diode current of less than, say 10%, the voltage drop due to any series resistance must be less than 2 mV. Any larger voltage drop is "significant".

A real diode can be modeled as two resistors and an ideal diode, as shown in Figure 7.9. The length of the p- region is W_P and the length of the n- region is W_N. R_P is the series resistance due to the p- region, and R_N is the series resistance due to the n- region. Considering only the majority carriers on each side, the two resistor values are:

$$R_P = \frac{W_P}{q A \mu_{pp} N_A} \tag{7.38}$$

$$R_N = \frac{W_N}{q A \mu_{nn} N_D} \tag{7.39}$$

where μ_{nn} is the electron mobility on the n-side and μ_{pp} is the hole mobility on the p-side. This is different that the minority carrier mobilities used earlier in this chapter. Keep in mind:

- Resistors are majority carrier devices.
- Diodes are minority carrier devices.

Equations 7.38 and 7.39 are only approximations. At high current densities, the minority carrier concentration can become high and reduce the resistance. And in the real world, there is contact resistance between the semiconductor and the metal contacts.

The voltage across the ideal diode (V_{pn}) is reduced from the voltage applied to the real diode (V_D) by the voltage drop across R_P and R_N. The voltage seen by the ideal diode is:

$$V_{pn} = V_D - IR_P - IR_N \tag{7.40}$$

Inserting the equation for diode current, Equation 7.40 becomes:

$$V_{pn} = V_D - I_S \exp\left(\frac{qV_{pn}}{k_BT}\right)(R_P + R_N) \tag{7.41}$$

This may be solved numerically for V_{pn}, and then the current may be found using equation 7.34.

In summary, there is always a voltage drop across R_P and R_N. When the current is small, such as reverse bias and a small forward bias, the voltage drop is negligible and equation 7.34 will accurately predict the current. When the current becomes large, the voltage drop across the two resistors becomes significant and the actual current is smaller than that predicted by Equation 7.34. This resistance prevents us from obtaining millions of amps of current by applying several Volt to a silicon diode.

Example 7.6: As a continuation of Example 7.5, consider a silicon diode with $N_A = 10^{15}$ cm^{-3} and $N_D = 10^{18}$ cm^{-3} and a cross-sectional area of 1 mm x 1 mm. The physical length of the p-region is 100 μm, and the physical length of the n-region is 10 μm. (a) What is the current for an applied voltage of 1 V if we ignore the series resistance? (b) Calculate R_N and R_P. (c) Calculate the current including the series resistance.

Answer: (a) In the last example, we found $J_S = 9.69$ pA/cm^2. The cross-sectional area is:

$A = 0.1$cm $\cdot$ 0.1cm $= 10^{-2}$cm^2 The saturation current is then:

$$I_S = A J_S = 10^{-2} \text{cm}^2 \cdot 9.69 \frac{\text{pA}}{\text{cm}^2} = 96.9 \text{ fA}$$

With 1 V applied, the current is:

$$I = I_S \left[\exp\left(\frac{qV_{pn}}{kT}\right) - 1 \right] = 96.9 \text{ fA} \left[\exp\left(\frac{1 \text{ V}}{0.0259 \text{ V}}\right) - 1 \right]$$

$$I = 5.68 \text{ kA}$$

That's a lot of current!

(b) R_N and R_P are solved using Equations 7.38 and 7.39. We have to find μ_{nn} and μ_{pp} from the tables in Appendix B.

$$\mu_{nn} = 261 \frac{\text{cm}^2}{\text{V} \cdot \text{s}}$$

$$\mu_{pp} = 458 \frac{\text{cm}^2}{\text{V} \cdot \text{s}}$$

$$R_P = \frac{W_P}{qA\mu_{pp}N_A} = \frac{100 \times 10^{-4}\text{cm}}{1.6 \times 10^{-19}\text{C} \cdot 10^{-2}\text{cm}^2 \cdot 458\frac{\text{cm}^2}{\text{V} \cdot \text{s}} \cdot 10^{15}\text{cm}^{-3}}$$

$$R_P = 13.5 \ \Omega \quad R_N = \frac{W_N}{qA\mu_{nn}N_D}$$

$$= \frac{10 \times 10^{-4}\text{cm}}{1.6 \times 10^{-19}\text{C} \cdot 10^{-2}\text{cm}^2 \cdot 261\frac{\text{cm}^2}{\text{V} \cdot \text{s}} \cdot 10^{18}\text{cm}^{-3}}$$

$$R_N = 0.002 \ \Omega$$

Notice that the lightly doped side determines the overall series resistance. The resistance on the heavily doped side is negligible.

(c) Let us ignore R_N because it is small compared to R_P. From Equation 7.41, we can write:

$$V_{pn} = 1V - 9.69 \text{ pA/cm}^2 \cdot 0.1\text{cm} \cdot 0.1\text{cm} \exp\left(\frac{qV_{pn}}{k_B T}\right)(13.5 \ \Omega)$$

This equation cannot be solved directly. Using a computer, we find $V_{pn} = 0.6792$ V.

The current can now be found using two methods. (1) We can find the current through the diode using Equation 7.34.

$$I = I_S \left[\exp\left(\frac{qV_{pn}}{kT}\right) - 1\right] = 96.9 \text{ fA} \left[\exp\left(\frac{0.6792 \text{ V}}{0.0259 \text{ V}}\right) - 1\right]$$

$$I = 23.7 \text{ } mA$$

(2) We can find the current through the resistor using Ohm's law:

$$I = \frac{V_{res}}{R} = \frac{1V - 0.6792V}{13.5 \text{ } \Omega} = 23.8 \text{ } mA$$

These two answers are essentially the same, except for a small rounding error. Notice that just $13.5 \text{ } \Omega$ changed the current from kA to mA.

7.6   Short Diodes

In our derivation for the diode current density, we calculated the slope of the excess carrier concentration at the edge of the depletion region, and then calculated the diffusion current. The calculation was performed by solving the continuity equation, Equation 7.24, and using the general solution, Equation 7.25. For boundary conditions, we used $\Delta p(x = \infty) = 0$ because we <u>assumed</u> that the length of the n- region was much greater than the diffusion length, L_p. <u>What if this assumption isn't valid?</u> After all, one of the goals in CMOS fabrication is to make things as small as possible.

Look at Figure 7.10, where the excess hole concentration in the n-region is plotted as a function of distance for two scenarios. In one scenario, the n- region is very long, and we treat it as semi-infinite. This is called a **long diode**. In the other scenario, the n- region is much smaller than the diffusion length. This type of diode is called a **short diode**.

Previously, the current density in a diode was solved for the long diode. To solve for the current density in a short diode, we need a new boundary condition to replace $\Delta p(x = \infty) = 0$. The new boundary condition is:

$$\Delta p(x = W_N) = 0 \tag{7.42}$$

Short Diode

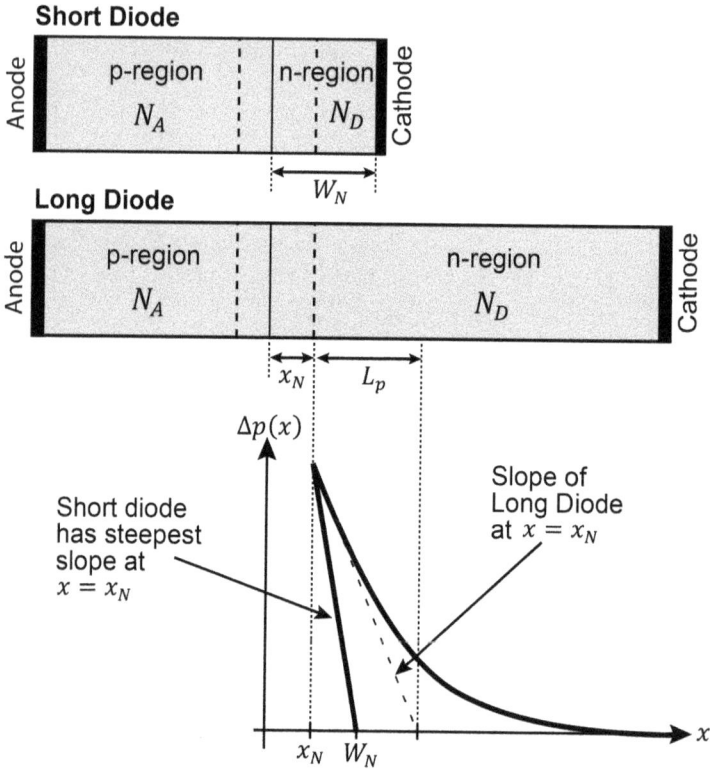

Figure 7.10: Excess hole concentration in the n- region for a long diode and a short diode. The slope of the excess hole concentration is shown for the long diode.

Unfortunately, solving this results in a very ugly equation. However, we may simplify the problem if $W_N - x_N \ll L_p$. Here, we can find an approximate solution by noticing that the slope of Δp is nearly linear, as seen in Figure 7.10. The diffusion current requires us to calculate $\frac{d\Delta p}{dx}$, which can be thought of as a "rise over run". For a long diode, the rise/run is seen in Figure 7.10 to be:

$$\frac{d\Delta p}{dx} = \frac{\text{Rise}}{\text{Run}}\bigg|_{\text{long}} = \frac{\Delta p(x = x_N)}{L_p} \tag{7.43}$$

And for a short diode, the rise/run is seen in Figure 7.8 to be:

$$\frac{d\Delta p}{dx} = \frac{\text{Rise}}{\text{Run}}\bigg|_{\text{short}} = \frac{\Delta p(x = x_N)}{W_N - x_N} \tag{7.44}$$

The only difference is the "run". Thus, we will replace the "run" in the current density equation for the long diode with the "run" for the short diode. Equation 7.32 had a "run" of L_p, but this should be replaced with $W_N - x_N$ for a short diode. Doing the same for the p- side, we can rewrite the equation for the saturation current density (Eq. 7.33) for the case that both the p- side and n- side are short diodes:

$$J_S = qn_i^2 \left(\frac{D_p}{(W_N - x_N)N_D} + \frac{D_n}{(W_P - x_P)N_A} \right) \tag{7.45}$$

If you get a diode in which one side is a long diode and the other side is a short diode, then you have to pick out the appropriate terms from Equations 7.33 and 7.45.

In summary, we have replaced a term in the denominator for the saturation current density with a smaller term. This makes the saturation current density larger. A short diode has a larger saturation current density, a larger reverse current density, and a larger forward current density than a long diode.

7.7 Breakdown Voltage

Breakdown in a diode is **non-destructive**. That's right, operating a diode past its breakdown point does not destroy the diode. So now you, the reader, are asking yourself, "Can I apply a million volts to a tiny diode and not destroy it?" Hah! I wish. So let's be a little more specific. Operating a diode past its breakdown point does not directly destroy the diode, but generating too much heat will destroy the diode.

What destroys a semiconductor device is temperature. If the chip gets too hot, it fails. It fails because something melts, or the plastic package burns. If the chip doesn't get too hot, then everything is fine. The amount of power that a chip can tolerate depends upon how it is packaged. If you put a heat sink on the device, it can tolerate more power. When you purchase a diode, it will typically rate the maximum power. This maximum power applies for both forward and reverse bias.

Breakdown in a p-n junction is a reverse bias phenomenon. In forward bias, the current is limited by the series resistance and the maximum power that may be dissipated. Reverse bias is a bit more interesting. Figure 7.11 shows a typical curve for a p-n junction in reverse bias. For

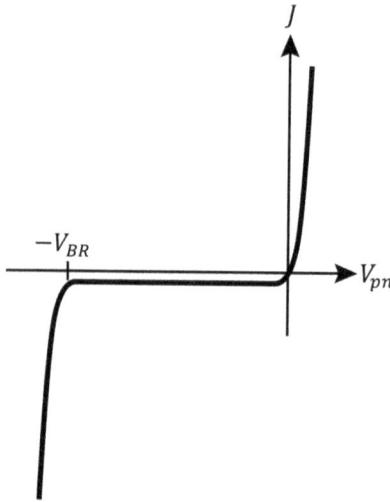

Figure 7.11: Typical I-V curve for a p-n junction including the effects of breakdown.

a large range of reverse bias voltages, the current density is equal to $-J_S$, which is very small. Once the reverse bias reaches the breakdown voltage, the current density increases very rapidly. In fact, the I-V curve often looks to be nearly vertical. Operating the p-n junction in this range is not destructive so long as the power dissipation does not become too large.

Notice that we have defined the Breakdown Voltage as the magnitude of the applied voltage required for breakdown to occur. That is, V_{BR} is a positive number, even though a negative voltage is applied to the diode. $V_{BR} = -V_{pn}$

It is easy to exceed the maximum power dissipation rating on a diode when operating in breakdown because power is $P = IV$. Since we are in breakdown, the voltage is typically large. A moderate current can easily turn into a very large power dissipation.

Typically, manufacturers are not interested in making a diode with a specific breakdown voltage. A diode might be rated for 200 V, but the actual breakdown voltage may range from 300 – 500 V. After all, meeting tight tolerances costs money. A Zener diode is a diode just like every other diode, but it is manufactured to have a specific breakdown voltage (tighter tolerance, and thus costs more money). Zener diodes may be used in a DC power supply for voltage regulation or to generate a reference voltage.

Breakdown occurs in a p-n junction for one of three reasons: (1) electrons tunnel through the depletion region (Zener Breakdown), (2) electrons increase in number due to carrier ionization (Avalanche Breakdown), or (3) the depletion region reaches an electrode (Punch-Through). Let us consider each of these in turn.

7.7.1 Zener-Breakdown

Electrons can pass through a barrier via quantum tunneling if the barrier is only a few nm thick. The tunneling rate becomes almost zero for a barrier greater than 5 nm. Figure 7.12 shows a band diagram where electron tunneling can occur. Electrons can tunnel from the valence band, where there are many electrons, to the conduction band, where there are many empty quantum states available to accept the electrons. Electrons do not change energy as they tunnel. Initially, an electron is on one side of the barrier; then, suddenly, it is on the other side of the barrier.

To obtain a very thin barrier (< 5 nm) requires very high doping concentrations on both sides of the depletion layer, and the breakdown voltage will be in the range of a few volts (< 3 V). Most p-n junctions are not made this way, and hence most diodes do not have Zener breakdown. Despite its name, if you purchase a Zener diode with a breakdown voltage greater than 3V, it does not operate by Zener breakdown.

Reverse Bias

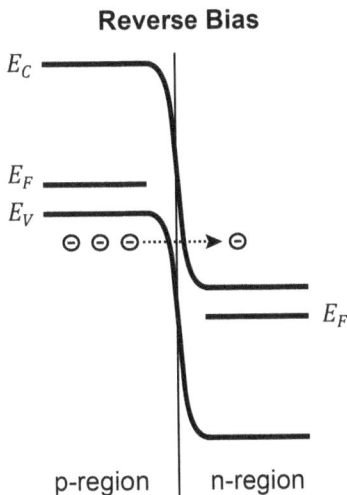

Figure 7.12: Band diagram showing electrons tunneling from the valence band to the conduction band under reverse bias conditions. This is called Zener breakdown.

Looking at Figure 7.12, we see that a negative voltage was applied to the p-region, moving the Fermi level moved up. Electrons tunnel from the p-region to the n-region, resulting in current flow from the n-region to the p-region. Using our convention, this is a negative current. Since there are a lot of electrons in a valence band, there is very little resistance to the current flow. Hence, under reverse bias, when Zener breakdown occurs, the magnitude of the current increases exponentially, and the current is limited by the series resistance of the circuit.

7.7.2   Avalanche Breakdown

Consider an electron in the depletion region with a large reverse bias applied, as shown in step 1 of Figure 7.13. There is a high electric field present because of the reverse bias, and the electric field accelerates the electron to the right. Eventually the electron runs into something, giving off its energy and slowing down. If the electric field is large enough, the electron can gain more energy than the band gap of the semiconductor, enabling the electron to release enough energy to move an electron from the valence band to the conduction band. See step 2 in Figure 7.13. Now there are two electrons in the conduction band and one hole in the valence band, or three mobile charges that carry current when we started with just one. But there's more! The two electrons can each ionize another pair of electrons and holes, and the hole moving in the other direction can ionize another electron/hole pair, as shown in

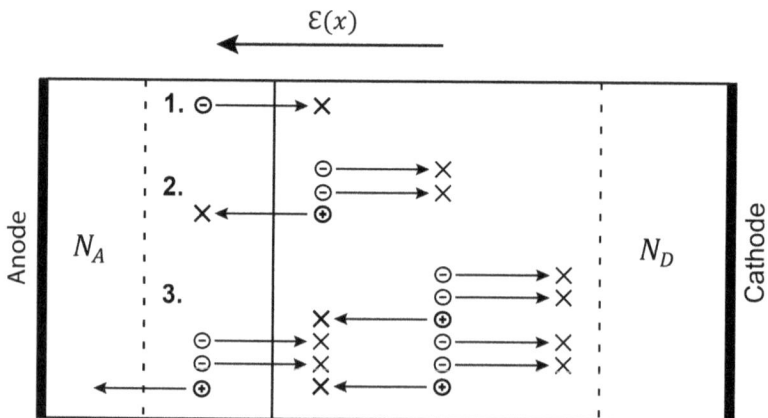

Figure 7.13: Schematic of the carrier multiplication effect that occurs during avalanche breakdown. In three steps, one electron becomes 6 electrons and 3 holes.

step 3 of Figure 7.13. There are now six electrons and three holes carrying current, meaning that the current has multiplied by a factor of 9 thus far. This multiplicative factor keeps increasing, and the current increases very rapidly in avalanche breakdown.

The term Avalanche Breakdown comes from a corollary with a snowball rolling down from the top of a mountain. It gains speed due to gravity, then it can hit some other snow and become two snowballs rolling down the mountain. These two snowballs hit two more, growing very rapidly into an avalanche down the side of the mountain.

A simple model for avalanche breakdown assumes that the p-n junction undergoes avalanche breakdown when the maximum electric field exceeds a critical electric field. The critical electric field for several semiconductors is shown in Table 7.1. The breakdown voltage may be calculated using Equation 7.15 and setting the electric field equal to the critical electric field. Then we solve for the voltage. The breakdown voltage is:

$$V_{BR} = -V_{pn} = \mathcal{E}_{crit}^2 \frac{\epsilon_s \epsilon_0}{2q} \left(\frac{N_D + N_A}{N_A N_D} \right) - \phi_{bi} \tag{7.46}$$

From an examination of Equation 7.46, several conclusions may be made. First, the breakdown voltage is determined by the lightly doped side of the junction. Second, the lighter the doping concentration, the greater the breakdown voltage. Third, the breakdown voltage varies as

Table 7.1: Critical Electric Field for avalanche breakdown.

Semiconductor	$\mathcal{E}_{crit}\left(\dfrac{V}{cm}\right)$
Silicon	3×10^5
GaAs	4×10^5
SiC	2×10^6
GaN	3×10^6
Diamond	10^7

the square of the critical electric field. Thus, diamond should have a breakdown voltage more than 1000x larger than a similar silicon diode. This motivates a lot of research on semiconductors such as SiC and GaN to replace silicon as a semiconductor for power electronics.

Example 7.7: Consider a diode made in silicon with a lightly doped n-region with $N_D = 10^{16}$ cm^{-3}. What is the breakdown voltage due to avalanche breakdown?

Answer: The breakdown voltage is given by Equation 7.46. If we assume that ϕ_{bi} is small and may be neglected, and assuming that the p- region is doped much greater than the n- region ($N_A \gg N_D$), we may approximate the breakdown voltage as:

$$V_{BR} = \mathcal{E}_{crit}^2 \frac{\epsilon_s \epsilon_0}{2q N_D}$$

Thus, the breakdown voltage is:

$$V_{BR} = (3 \times 10^5 \text{ V/cm})^2 \frac{11.8 \cdot 8.854 \times 10^{-14} \text{ F/cm}}{2 \cdot 1.6 \times 10^{-19} \text{ C} \cdot 10^{16} \text{ cm}^{-3}} = 29 \text{ V}$$

Example 7.8: Consider a diode made in silicon with a lightly doped p-region with $N_A = 10^{15}$ cm^{-3}. What is the breakdown voltage due to avalanche breakdown?

Answer: The breakdown voltage is given by Equation 7.46. If we assume that ϕ_{bi} is small and may be neglected, and assuming that the n- region is doped much greater than the p- region ($N_D \gg N_A$), we may approximate the breakdown voltage as:

$$V_{BR} = \mathcal{E}_{crit}^2 \frac{\epsilon_s \epsilon_0}{2q N_A}$$

Notice that this is the same equation that we found in Example 7.7, but with N_A replacing N_D. That is, the lightly doped region determines the breakdown voltage. The breakdown voltage is:

$$V_{BR} = (3 \times 10^5 \text{ V/cm})^2 \frac{11.8 \cdot 8.854 \times 10^{-14} \text{ F/cm}}{2 \cdot 1.6 \times 10^{-19} \text{ C} \cdot 10^{15} \text{ cm}^{-3}} = 290 \text{ V}$$

This breakdown voltage is much larger than that found in Example 7.7. This example shows that it is necessary to have a very lightly doped region in order to sustain a large breakdown voltage.

7.7.3   Punch-Through

Punch-through occurs when the depletion region reaches a boundary and cannot grow any further. For example, punch-through occurs if the depletion region reaches the anode or cathode electrode. This occurs most often in a short diode. To determine if punch-through will occur, calculate the extent of the depletion region in the n-region (x_N) and the p-region (x_P) and compare it to the physical length available.

Example 7.9: As a continuation of Example 7.5, consider a silicon diode with $N_A = 10^{15}$ cm^{-3} and $N_D = 10^{18}$ cm^{-3}. The length of the lightly doped p-region is 10 µm. What is the breakdown voltage due to punch-through?

Answer: Most of the depletion region extends into the lightly doped region, so this is typically where punch-through occurs. The breakdown voltage due to punch-through occurs when the depletion region width in the p- region is equal to 10 µm. From Example 7.2, we know that the depletion region width is 1 µm without an applied voltage, and the built-in potential is 0.78 V. The equation for x_P is:

$$x_P = \sqrt{\frac{2\epsilon_s\epsilon_0}{qN_A}\left(\frac{N_D}{N_D + N_A}\right)\left(\phi_{bi} - V_{pn}\right)}$$

Solving for the applied voltage, the breakdown voltage is $V_{BR} = -V_{pn}$:

$$V_{BR} = -V_{pn} = \frac{qN_A}{2\epsilon_s\epsilon_0}\left(\frac{N_D + N_A}{N_D}\right)x_P^2 - \phi_{bi}$$

Inserting numbers:

$$V_{BR} = \frac{1.6 \times 10^{-19}\,\text{C} \cdot 10^{15}\,\text{cm}^{-3}}{2 \cdot 11.8 \cdot 8.854 \times 10^{14}\,\text{F/cm}}\left(\frac{10^{18}\,\text{cm}^{-3} + 10^{15}\,\text{cm}^{-3}}{10^{18}\,\text{cm}^{-3}}\right)(10$$
$$\times 10^{-4}\,\text{cm})^2 - 0.78\,\text{V}$$

The length of the p- region was converted to cm to permit the units to cancel. The breakdown voltage due to punch-through is:

$$V_{BR} = 76\,\text{V}$$

This is smaller than the breakdown voltage calculated in Example 7.8. Punch-through will occur before avalanche breakdown.

7.8   High-Level Injection

In section 7.3, we used an approximation called low-level injection to derive the electron and hole concentrations in the p-n junction. Low-level injection is an approximation that says that the majority carrier concentration does not significantly change, and that only the minority carrier concentration will change with an applied voltage. In this section, we will see what happens if we relax that approximation.

If the applied voltage is sufficiently large, then the minority carrier concentration will become larger than doping concentration. This is called **high level injection**. The following example shows how this can arise in practice.

Example 7.10: Consider a silicon diode with $N_A = 10^{15}$ cm^{-3} and $N_D = 10^{18}$ cm^{-3}.   What   are   the   excess   minority   carrier concentrations next to the depletion region for an applied voltage of 0.6 V?

Answer: For an applied voltage of 0.7 V, which corresponds to forward bias:

$$\Delta p(x = x_N) = \frac{n_i^2}{N_D}\left[\exp\left(\frac{qV_{pn}}{k_BT}\right) - 1\right]$$

$$\Delta p(x = x_N) = \frac{10^{20}\ cm^{-6}}{10^{18}\ cm^{-3}}\left[\exp\left(\frac{0.7\ V}{0.0259\ V}\right) - 1\right] = 5.5 \times 10^{13}\ cm^{-3}$$

$$\Delta n(x = -x_P) = \frac{n_i^2}{N_A}\left[\exp\left(\frac{qV_{pn}}{k_BT}\right) - 1\right]$$

$$\Delta n(x = -x_P) = \frac{10^{20}\ cm^{-6}}{10^{15}\ cm^{-3}}\left[\exp\left(\frac{0.7\ V}{0.0259\ V}\right) - 1\right] = 5.5 \times 10^{16}\ cm^{-3}$$

In the n-region, the excess minority carrier concentration is far below the doping concentration: $\Delta p \ll N_D$. In the p-region, the electron concentration is $n \approx \Delta n$. This is greater than N_A, and therefore the minority carrier concentration exceeds the doping concentration.

One thing to note is that high-level injection can only occur in forward bias. In reverse bias, the carrier concentrations are reduced near the junction, and therefore the carrier concentration will not approach the doping concentration.

The second thing to note is that when high-level injection occurs, charge neutrality will force the electron and hole concentrations to be approximately equal. This comes about from Equation 4.10:

$$\rho(x) = q[p(x) - n(x) + N_D^+(x) - N_A^-(x)] = 0 \qquad (7.47)$$

This equation is equal to zero because we are considering the neutral region of the diode. Let us consider an n-type semiconductor: $N_D > 0$ and $N_A = 0$. In high level injection, the minority carrier concentration, p, is greater than the doping concentration. That is, $p \gg N_D$. Thus, N_D is negligible and may be ignored in Equation 7.47. The only way to get $\rho = 0$ is if $p \approx n$.

The derivation for $\Delta n(x)$ and $\Delta p(x)$ that we performed in section 7.3 is not valid for high-level injection. Therefore, a new equation for current in high-level injection needs to be derived. The overall result is that the diode current increases slower with the application of a voltage than in low-level injection. Since the current is limited at high voltage by high-level injection, and the current is limited at high voltage due to series resistance, in practice it is often not clear which effect is limiting the current.

Some semiconductor devices deliberately use high-level injection. For example, a **p-i-n diode** uses a lightly doped region to obtain a high breakdown voltage. Normally, this would result in a high series resistance, but by operating in high-level injection, the carrier concentration is high and the series resistance drops. A second type of device that uses high-level injection is the **Insulated Gate Bipolar Transistor (IGBT)**. This device is described in Chapter 13.

7.9   Temperature Sensitivity

The diode is very sensitive to temperature. We will look at forward and reverse bias separately.

In reverse bias, we need to find the temperature dependence of the saturation current density. Although there are terms such as the diffusion coefficients, D_n and D_p, that do change with temperature, we will ignore this effect because the temperature effect on the intrinsic carrier concentration will dominate. From Equation 7.33, we find:

$$\frac{dJ_S}{dT} = 2qn_i \left(\frac{D_p}{L_p N_D} + \frac{D_n}{L_n N_A} \right) \frac{dn_i}{dT} \qquad (7.48)$$

The term $\frac{dn_i}{dT}$ can be calculated from Equation 3.37. Conceptually, notice that the $\frac{dn_i}{dT}$ term is very large and positive, as shown in Figure 3.13. Therefore, the saturation current density increases with temperature, and the magnitude of the current density increases with temperature in reverse bias.

The forward bias situation is more complicated because there are two terms with an exponential change with respect to temperature. As seen in Equation 7.48, the saturation current (J_S) causes the diode current to increase. In contrast, the voltage exponential term, $\exp\left(\frac{qV_{pn}}{k_B T}\right)$, works to decrease the diode current as the temperature increases. In addition, we need to consider the series resistance. The series resistance (R_N and R_P) will decrease with temperature, causing more of the diode voltage V_D to be applied to V_{pn}. This leads to an increase in current. In practice, when all effects are taken into account, the current in a diode will tend to increase as the temperature increases.

In summary, as the temperature increases:

- In forward bias, the current density increases.
- In forward bias, the perceived turn-on voltage decreases.
- In reverse bias, the leakage current increases.

7.10 Summary of Key Concepts

- Current is carried by free carriers: electrons and holes.
- Dopant atoms, such as acceptors and donors, are fixed charges and cannot carry current.
- There is a depletion region at the interface between the p- region and the n- region. Fixed charges, due to dopant atoms, dominate the charge density in the depletion region and determine the electric field and electric potential.
- Despite the name 'depletion', in a depletion region there are still electrons and holes. It is just that the concentration of electrons

and holes is substantially smaller than the concentration of dopant atoms.

- Because there are electrons and holes in the depletion region, current can flow through a depletion region under the right conditions.

- In equilibrium, there is an electric potential difference from the p- region to the n- region due to the charges in the depletion region. This electric potential difference is called the built-in potential, or ϕ_{bi}.

- The electric field points from the n- region to the p- region. Hence, the electric field is negative.

- The quantity of charge in the depletion region on the n-side must equal the quantity of charge on the p-side.

- The current in a p-n junction is determined by the minority carriers: the electrons in the p- region and the holes in the n-region.

- With zero voltage applied, the drift current in the depletion region is exactly balanced by the diffusion current, resulting in zero net current. A drift current must exist because of the electric field in the depletion region.

- By applying a positive voltage to the p-n junction, the device is forward biased.

- By applying a negative bias to the p-n junction, the device is reverse biased.

- In forward bias, the electric potential between the p- region and n- region is reduced; the magnitude of the electric field is reduced; and the depletion layer width is reduced.

- In reverse bias, the electric potential between the p- region and n- region is increased; the magnitude of the electric field is increased; and the depletion layer width is increased.

- In forward bias, the excess carrier concentrations are positive. In reverse bias, the excess carrier concentrations are negative.

- In forward bias, the current density increases exponentially with applied voltage.

- In reverse bias, the current density very quickly saturates at $-J_S$, which is a very small number. Thus, it is commonly said that a diode blocks all current in reverse bias.

- Semiconductors are not perfect conductors. The series resistance within the n- region and p- region can significantly reduce the current through a p-n junction.
- A short diode is one in which the excess carrier concentration does not reduce to zero prior to reaching the edge of the semiconductor. Rather, the excess carrier concentration is forced to zero at the edge of the semiconductor. The reverse saturation current density of a short diode is greater than that of a similarly doped long diode.
- Breakdown is non-destructive, unless the semiconductor is damaged by heat.
- Breakdown may occur due to Zener breakdown, Avalanche breakdown, or punch-through.
- A semiconductor is in low level injection when the minority carrier concentration is smaller than the doping concentration.
- A semiconductor is in high level injection when the minority carrier concentration is larger than the doping concentration.
- The current density in a diode increases as the temperature goes up. This is primarily due to the temperature dependence of n_i.

7.11 Problems

1. Draw the energy band diagram of a p-n junction, the charge density, the electric field, and the electric potential for in equilibrium.

2. Draw the energy band diagram of a p-n junction in (a) Forward Bias and (b) Reverse Bias.

3. What is the built-in electric potential for a p-n junction with $N_A = 10^{17}$ cm^{-3} and $N_D = 10^{18}$ cm^{-3} for:
   a. Silicon
   b. GaN

4. What is the extent of the depletion region in the p- region and the n- region with $N_A = 10^{17}$ cm^{-3} and $N_D = 10^{18}$ cm^{-3} in equilibrium for:
   a. Silicon
   b. GaN

5. What is the maximum electric field for a p-n junction with $N_A = 10^{17}$ cm^{-3} and $N_D = 10^{18}$ cm^{-3} in equilibrium for:
   a. Silicon
   b. GaN

6. Consider a silicon p-n junction with $N_A < N_D$. Let us look at the energy bands and electron concentration within the p-region.

   a. Sketch the energy band diagram in the p-region. Then sketch $E_{iF}(x) = E_i - E_F$.
   b. Derive an equation for $E_i - E_F$ as a function of x, for $-x_p < x < 0$ (within the depletion region on the p-side). That is, find $E_{iF}(x)$. Start with $E_i - E_F$ at $x = -x_p$, and add $-q\phi(x)$. (Energy bands are the mirror image of potential.) Use Equation 7.9.
   c. If $N_A = N_D$, what is $E_{iF}(0)$?
   d. Use a computer, and plot $E_{iF}(x)$ over the range $-x_p < x < 0$ for $N_A = 10^{15}$ cm^{-3} and $N_D = 5 \times 10^{15}$ cm^{-3}. The graph should have the same shape as the energy bands in part (a).
   Note: In part (d), you should notice that E_{iF} becomes negative. That is, there are more electrons than holes near $x = 0$. (i.e., the p-n junction is in invesion).
   e. Use a computer, and plot $n(x)$ over the range $-x_p < x < 0$ for $N_A = 10^{15}$ cm^{-3} and $N_D = 5 \times 10^{15}$ cm^{-3}.
   f. For part (e), what is the maximum electron concentration? Is this in weak inversion, or strong inversion?
   g. Repeat parts (d) and (e) for $N_A = 10^{15}$ cm^{-3} and $N_D = 10^{17}$ cm^{-3}.
   h. For part (g), what is the maximum electron concentration? Is this in weak inversion, or strong inversion?
   Note: You should have found that the p-region is in strong inversion. That means the depletion approximation is not valid in this region!

7. Consider a silicon p-n junction with $N_A = 10^{17}$ cm^{-3} and $N_D = 10^{17}$ cm^{-3}. The minority carrier lifetime on the p-side is 10 µs, and the minority carrier lifetime on the n-side is 100 µs.
   a. What is the built-in potential, ϕ_{bi}?
   b. What is the excess electron concentration at $x = -x_p$, for $V_{pn} = -3\ V$?
   c. What is the excess electron concentration at $x = -x_p$, for $V_{pn} = 0.5\ V$?
   d. What is the reverse saturation current density, J_S?
   e. What is the current density for $V_{pn} = -3\ V$?
   f. What is the current density for $V_{pn} = 0.5\ V$?

8. Consider a GaN p-n junction with $N_A = 10^{18}$ cm^{-3} and $N_D = 10^{15}$ cm^{-3}. The minority carrier lifetime on the p-side is 10 ns, and the minority carrier lifetime on the n-side is 10 µs. The cross-sectional area is 100 µm x 100 µm. The length of the n-region is 200 µm, and the length of the p- region is 10 µm. What are:
   a. The built-in potential, ϕ_{bi}.
   b. The resistance of the n- region.
   c. The resistance of the p- region.
   d. The breakdown voltage considering only avalanche breakdown.
   e. The breakdown voltage considering only punch-through.
   f. The breakdown voltage.

9. Consider a GaAs p-n junction with $N_A = 10^{18}$ cm^{-3} and $N_D = 10^{16}$ cm^{-3}. The minority carrier lifetime on the p-side is 10 ns, and the minority carrier lifetime on the n-side is 1 µs.
   a. What is the reverse saturation current density, J_S?
   b. Using a computer, plot the current density versus applied voltage, ranging from -2 V to 1 V.

   Note: You should play around with the maximum voltage that you plot. You should be able to vary the turn-on voltage.

10. It is desired to make a very thin diode. The p-side of the diode will be very heavily doped compared to the n- side. Determine the correct doping concentration on the n-side (N_D) to achieve a depletion width inside the n-side of 10 nm. (Note: This type of calculation is important for the design of very small MOSFETs.)

11. When we discuss how small we can make a MOSFET, we will learn that one limitation is from two reverse biased diodes near each other (source and drain), and they should not interact. Use $N_D = 10^{20} cm^{-3}$. Assume that $N_D \gg N_A$. Make a plot of $2W_D$ versus doping concentration (N_A) for:
    a. Zero applied voltage
    b. Zero applied voltage on one diode and 3.3 V reverse voltage on the other
    c. Zero applied voltage on one diode and 5 V reverse voltage on the other

    Use a log-log plot, and use a computer. For the x-axis, use a range from $N_A = 10^{14}$ cm^{-3} to $N_A = 10^{19}$ cm^{-3}.

    d. Modern MOSFETs have a gate length below 50 nm. How large must the doping concentration be to keep $2W_D$ below 50 nm?

12. A silicon p$^+$n junction has $N_A = 10^{19}$ cm^{-3} and $N_D = 2 \times 10^{16}$ cm^{-3}. The cross-sectional area is 0.5 mm x 0.5 mm, $\tau_p = 10$ μs, and $D_p = 10$ cm^2/s. The length of the p-region is 5 μm, and the length of the n-region is 500 μm.
    a. What is the contact potential of the pn junction?
    b. What is the series resistance of the p-region?
    c. What is the series resistance of the n-region?
    d. What is the total series resistance of the diode?
    e. What is the reverse saturation current, I_S?
    f. Plot the current as a function of voltage, for V=0 .. 1 V. Include the effects of series resistance. You must use a computer to solve the resulting implicit equation. You cannot find a simple equation for the current versus voltage.
    g. On the same graph, plot the current as a function of voltage, for V=0 .. 1 V. Do not include the effects of series resistance.

h. Make a new plot with two traces, as per parts (f) and (g). This time use a log scale on the y-axis. Over what voltage range does the series resistance have a significant impact on the current flow?

i. How much power is dissipated at 1V forward bias?

j. A glass encapsulated diode (a typical package) has a thermal resistance (junction to ambient) of 50 °C/W. What is the steady state temperature for the diode at 1 V forward bias, including the series resistance? Hint: $\Delta T = R_{th}P_{diss}$

13. For the diode described in problem #12, what are:
   a. The breakdown voltage due to avalanche breakdown?
   b. The breakdown voltage due to punch-through?
   c. The actual breakdown voltage (minimum of parts (a) and (b))?

14. (Hard) In section 7.6 we discussed short diodes. We used the assumption that $W_N - x_N \ll L_p$ and $W_P - x_P \ll L_n$ in order to obtain a linear slope for Δp and Δn. In this problem, let us find an exact equation for the current in a p-n junction that is valid whether the diode is short, long, or something in-between.

   a. Calculate the excess hole concentration in the n- region using Equation 7.23 for one boundary condition and the following boundary condition: (The answer should be simplified to contain hyperbolic functions)
   $$\Delta p(x = W_N) = 0$$

   b. Calculate the excess electron concentration in the p-region using Equation 7.27 for one boundary condition and the following boundary condition: (The answer should be simplified to contain hyperbolic functions)
   $$\Delta n(x = -W_P) = 0$$

   c. Find the hole diffusion current density at $x = x_N$ by differentiating your answer to part (a) using equation 6.5.

   d. Find the electron diffusion current density at $x = -x_P$ by differentiating your answer to part (b) using equation 6.4.

   e. Add your answers from parts (c) and (d) to obtain the total current density.

15. Consider a GaAs p-n junction with $N_A = 10^{17}$ cm^{-3} and $N_D = 10^{15}$ cm^{-3}. The minority carrier lifetime on the p-side is 100 ns, and the minority carrier lifetime on the n-side is 10 µs. For this problem, you have to calculate n_i as a function of temperature. See section 3.13.1.

    a. What is the built-in potential, ϕ_{bi}, at 300 K?

    b. What is the built-in potential, ϕ_{bi}, at 400 K?

    c. Does the built-in potential become larger or smaller with an increase in temperature?

    d. What is the depletion layer width, W_D, at 300 K?

    e. What is the depletion layer width, W_D, at 400 K?

    f. Does the depletion layer width become larger or smaller with an increase in temperature?

    g. What is the reverse saturation current density, J_S, at 300 K?

    h. What is the reverse saturation current density, J_S, at 400 K?

    i. Does the reverse saturation current density become larger or smaller with an increase in temperature?

8 Solar Cells and Photodiodes

In this chapter, we will learn how light can be absorbed in a semiconductor, generating both electrons and holes. The electrons and holes change the semiconductor resistance. The electrons and holes also contribute to the current in the p-n junction. Finally, we will see how a p-n junction becomes a solar cell.

8.1 Light Absorption in a Semiconductor

Light exists as both a wave and a particle. The name of a particle of light is a **photon**. Light, of course, comes in multiple colors, or wavelengths. Blue light has a wavelength of approximately 400 nm, and red light as a wavelength of approximately 700 nm. The **energy of a single photon** is:

$$E_{ph} = h\nu = \hbar\omega = \frac{hc}{\lambda} \qquad (8.1)$$

where $h = 6.626 \times 10^{-34}$ J·s is Plank's constant, $\hbar = \frac{h}{2\pi}$ is the reduced Plank's constant, ω is the angular frequency of light in radians/s, ν is the frequency of light in Hz, $c = 3 \times 10^8$ m/s is the speed of light, and λ is the wavelength of light in meters. A useful relation to find the **photon energy** in eV, from the wavelength in μm, is:

$$E_{ph}(eV) = \frac{1.24 \ (eV \cdot \mu m)}{\lambda} \qquad (8.2)$$

When a photon of energy E is absorbed, conservation of energy requires that an electron gain the same energy. Since most of the electrons are in the valence band, and very few electrons are in the conduction band, this means that 99.999% of the time (or more!) the electron that absorbs the photon is located in the valence band. Consider Figure 8.1. If an electron in the valence band tries to absorb a photon with an energy less than the band gap, then the electron is going to try to occupy a quantum state within the band gap, which does not exist. Therefore,

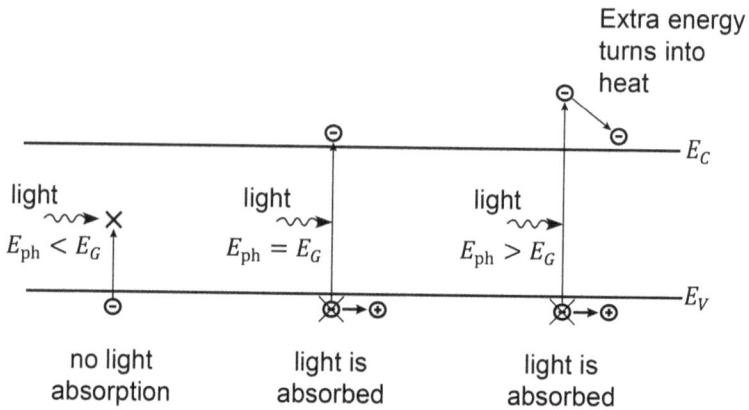

Figure 8.1: A photon may or not be absorbed by a semiconductor, depending on its wavelength. On the left, the photon cannot be absorbed because the optical energy is smaller than the band gap. The electron cannot occupy a state within the band gap. In the middle example, light is absorbed, moving an electron from the valence band to the conduction band, leaving behind a hole in the valence band. On the right, a photon with an energy greater than the band gap is absorbed. The extra energy is converted to heat after a very short amount of time.

semiconductors cannot absorb photons with energy less than the band gap energy. On the other hand, if an electron in the valence band tries to absorb a photon with an energy greater than the band gap, then that electron will move to the conduction band (not necessarily at E_C), which is permitted. Then the electron will rapidly lose energy as heat and settle down to an energy equal to the bottom of the conduction band (E_C). Therefore, the requirement for light absorption is:

$$E_{ph} \geq E_G \quad \text{(requirement for light absorption)} \tag{8.3}$$

When a photon is absorbed, an electron moves from the valence band to the conduction band. That is, one hole is generated, **and** one electron is generated. This is called an electron-hole pair.

Example 8.1: The visible light spectrum ranges from 400 nm to 700 nm. What is the energy of a photon with a wavelength of (a) 400 nm and (b) 700 nm?

Answer:
(a) Using Equation 8.2:
$$E_{ph} = \frac{1.24 \ (eV \cdot \mu m)}{\lambda} = \frac{1.24 \ (eV \cdot \mu m)}{0.4 \ \mu m} = 3.1 \ eV$$
Notice that we converted the wavelength to μm in order to use Equation 8.2.
(b)
$$E_{ph} = \frac{1.24 \ (eV \cdot \mu m)}{\lambda} = \frac{1.24 \ (eV \cdot \mu m)}{0.7 \ \mu m} = 1.77 \ eV$$

In the visible light spectrum, blue light (400 nm) has the most energy. In Example 8.1 we found that blue light has an energy of 3.1 eV. Therefore, if the band gap of a semiconductor is greater than 3.1 eV, then it cannot absorb any visible light and must be **transparent**. Looking at Appendix A, we see that both GaN and SiC have band gaps greater than 3.1 eV, and they are indeed as transparent as glass. Ge, GaAs, and silicon all have a band gap less than 3.1 eV, and they are **opaque**, although if these semiconductors have a mirror polished surface they will reflect a lot of light due to their high dielectric constant.

Silicon is transparent to most of the IR spectrum because silicon has a band gap of 1.12 eV and therefore transparent to photon wavelengths greater than 1.1 μm.

There are multiple methods of representing light intensity. In this book, **light intensity** will be indicated in terms of power per area. In this form, light intensity is also called **optical power.** Since the units are in terms of power, the symbol P_{ph} will be used, and the units are W/cm². The subscript 'ph' stands for photon. Sometimes the light is made of a single wavelength, such as from a laser. Most commonly, light is made up of many wavelengths, such as visible light made up of all colors of the rainbow. If light is made up of many wavelengths, with varying power for each wavelength, then the total light intensity must be found by integrating the optical power over all wavelengths:

$$P_{ph} = \int P'_{ph} \, d\lambda \tag{8.4}$$

This makes problems with multiple wavelengths very complicated. Most of the problems in this book will use monochromatic light for ease of understanding.

Example 8.2: Light shines from an LED with a wavelength of 500 nm and an optical power of $P_{ph} = 1 \, \text{mW/cm}^2$. What is the equivalent number of photons per second per cm^2?

Answer: First, we need to find E_{ph} using Equation 8.2:

$$E_{ph} = \frac{1.24}{\lambda} = \frac{1.24}{0.5 \, \mu\text{m}} = 2.48 \, \text{eV}$$

Next, we divide the light intensity by the energy of a photon:

$$\frac{P_{ph}}{E_{ph}} = \frac{10^{-3} \, \text{W/cm}^2}{2.48 \times 1.6 \times 10^{-19} \, \text{J}} = 2.5 \times 10^{15} \, \frac{\text{photons}}{\text{cm}^2 \cdot \text{s}}$$

Notice that we had to convert the photon energy to Joules to get the units to cancel. $1 \, \text{W} = 1 \, \text{J/s}$.

Light is not absorbed just at the surface of a material. Light is absorbed over a distance. The light intensity decays exponentially as the photons are absorbed, as shown in Figure 8.2. The following equation describes the light intensity as a function of distance:

$$P_{ph}(x) = P_{ph,0} \exp\left(-\frac{x}{L_A}\right) = P_{ph,0} \exp(-\alpha x) \tag{8.5}$$

where $P_{ph,0}$ is the light intensity at the semiconductor surface, L_A is the absorption length with units of cm, and α is the absorption coefficient with units of 1/cm. Equation 8.5 only works for monochromatic light because the absorption length and absorption coefficient vary with wavelength. The absorption coefficient is the inverse of absorption length: $\alpha = 1/L_A$.

The rate at which electron-hole pairs are **generated** is equal to the rate at which photons are absorbed. The **generation rate** of electrons and holes is found by (1) differentiating Equation 8.5 to get the rate of change of photons (which is negative), (2) dividing by the energy of a photon to convert from power (W) to number of photons per second,

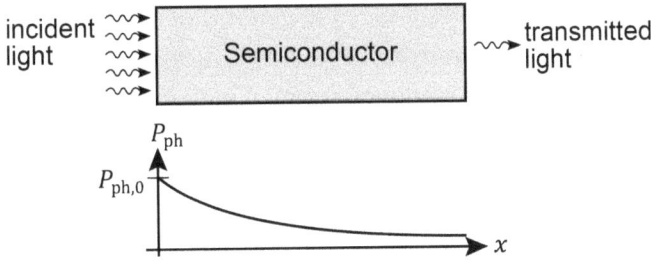

Figure 8.2: Light is incident on the semiconductor on the front (left side, x=0). As the light transmits through the semiconductor, the light is absorbed and the intensity drops. At the back side of the semiconductor, the remaining light that was not absorbed is transmitted. Any optical reflections are ignored in this illustration.

and (3) multiplying by negative one to convert from the rate of photons absorbed to the rate at which electrons and holes are generated. The generation rate is:

$$G(x) = \frac{P_{ph,0}}{E_{ph}L_A} \exp\left(-\frac{x}{L_A}\right) = \frac{\alpha P_{ph,0}}{E_{ph}} \exp(-\alpha x) \tag{8.6}$$

The generation rate, G, can be used in the continuity equation. The **generation rate** has units of #electrons/cm^3/s, or $1/(\text{cm}^3 \cdot \text{s})$. Note that Equation 8.6 only works for monochromatic light.

8.2 Resistance Change in a Semiconductor with Light

When light shines on a semiconductor, the concentration of electrons and holes changes. This, in turn, changes the conductivity, or resistivity, of the semiconductor. Since the light causes an increase in the electron and hole concentration, it is expected that the conductivity will increase and the resistivity will decrease. Using the continuity equations in steady state, and assuming zero current flow, the following equations are obtained:

$$0 = 0 - \frac{\Delta n}{\tau} + G \tag{8.7}$$

$$0 = -0 - \frac{\Delta p}{\tau} + G \tag{8.8}$$

For simplicity, let us assume that the generation rate is independent of x. This can occur, for example, if the absorption length is much greater than the thickness of the semiconductor. Then the generation rate at the top of the semiconductor will be nearly the same as the generation rate at the bottom of the semiconductor. Using equation 8.6, with $x = 0$, the generation rate is:

$$G = \frac{P_{ph,0}}{E_{ph}L_A} \tag{8.9}$$

The continuity equations can be solved for Δn and Δp:

$$\Delta n = \Delta p = \tau G = \tau \frac{P_{ph,0}}{E_{ph}L_A} \tag{8.10}$$

This shows that the excess carrier concentration is proportional to the generation rate of electrons and holes. It also shows that a longer carrier lifetime results in a larger excess carrier concentration.

Since $n = n_0 + \Delta n$ and $p = p_0 + \Delta p$, the conductivity of a semiconductor can be written as:

$$\begin{aligned} \sigma &= q(\mu_n n + \mu_p p) \\ &= q(\mu_n n_0 + \mu_p p_0 + \mu_n \tau G + \mu_p \tau G) \end{aligned} \tag{8.11}$$

The change in conductivity is:

$$\Delta\sigma = \sigma - \sigma_0 = q(\mu_n + \mu_p)\tau G \tag{8.12}$$

where σ_0 is the conductivity with no light. This shows that light causes the conductivity to increase. For low levels of light intensity, the excess carrier concentration is small compared to the majority carrier concentration, and the change in conductivity will be very small. For high levels of light intensity, the change in conductivity can be quite large. If there are many defects in the semiconductor, then the carrier lifetime will be short, resulting in a small change in conductivity. To achieve the highest sensitivity, it is desired to use a semiconductor with a long lifetime, which requires a semiconductor with as few defects as possible.

The change in resistance, ΔR, can be calculated as follows:

$$\Delta R = R - R_0 = \frac{L}{\sigma WH} - \frac{L}{\sigma_0 WH} = \frac{L}{WH}\left(\frac{\sigma_0 - \sigma}{\sigma_0 \sigma}\right)$$

$$= -R_0\left(\frac{\sigma - \sigma_0}{\sigma}\right) \tag{8.13}$$

where R_0 is the resistance with no light. This result shows that the resistance becomes smaller due to light. Using Equation 8.11 and 8.12, ΔR may be expanded as follows:

$$\Delta R = -R_0\left(\frac{(\mu_n + \mu_p)\tau G}{\mu_n n_0 + \mu_p p_0 + (\mu_n + \mu_p)\tau G}\right) \tag{8.14}$$

Example 8.3: Consider a slab of silicon with $N_A = 10^{15}$ cm^{-3}. The carrier lifetime is 100 µs. Light with a wavelength of 550 nm shines on the semiconductor at a power of 100 µW/cm^2. The absorption length is 2 µm. What are (a) the equilibrium electron and hole concentrations, (b) the electron and hole concentrations with light, and (c) what is the change in resistance if the sample initially had a resistance of 100 Ω ?

Answer: The equilibrium electron and hole concentrations are:

$$p_0 = N_A = 10^{15} \text{ cm}^{-3}$$

$n_0 = \frac{n_i^2}{p_0} = \frac{10^{20} \text{ cm}^{-6}}{10^{15} \text{ cm}^{-3}} = 10^5$ cm^{-3} To calculate the excess electron and hole concentrations with light, we need the generation rate:

$$E_{ph}(eV) = \frac{1.24}{\lambda(\mu m)} = \frac{1.24}{0.55} = 2.25 \text{ eV}$$

$$G = \frac{P_{ph}}{E_{ph}L_A} = \frac{10^{-4} \text{ W/cm}^2}{2.25 \text{ eV}\cdot 2 \times 10^{-4} \text{ cm}} \times \frac{1 \text{ eV}}{1.6 \times 10^{-19} \text{ J}}$$
$$= 1.39 \times 10^{18} \text{ cm}^{-3}\text{s}^{-1}$$

(a) The excess electron and hole concentrations are:

$$\Delta n = \Delta p = \tau\frac{P_{ph,0}}{E_{ph}L_A} = 10^{-4} \text{ s} \cdot 1.39 \times 10^{18} \text{ cm}^{-3}\text{s}^{-1}$$

$$\Delta n = \Delta p = 1.39 \times 10^{14} \text{ cm}^{-3}$$

(b) The electron and hole concentrations are:

$$n = n_0 + \Delta n = 10^5 \text{ cm}^{-3} + 1.39 \times 10^{14} \text{ cm}^{-3}$$

$$n = 1.39 \times 10^{14} \text{ cm}^{-3}$$

$$p = p_0 + \Delta p = 10^{15} \text{ cm}^{-3} + 1.39 \times 10^{14} \text{ cm}^{-3}$$

$$p = 1.139 \times 10^{15} \text{ cm}^{-3}$$

This shows us that the electron concentration changes quite a lot. This is because the electrons are the minority carrier. The hole concentration does not change much, but it will be enough to change the resistance.

(c) The change in resistance is:

$$\Delta R = -R_0 \left(\frac{(\mu_n + \mu_p)\Delta n}{\mu_n n_0 + \mu_p p_0 + (\mu_n + \mu_p)\Delta n} \right)$$

$$= -100 \, \Omega \left(\frac{(1314 + 458) \cdot \frac{\text{cm}^2}{\text{V} \cdot \text{s}} \cdot 1.39 \times 10^{14} \text{ cm}^{-3}}{1314 \cdot 10^5 + 458 \cdot 10^{15} + (1314 + 458) \cdot 1.39 \times 10^{14}} \right)$$

$$\Delta R = -25 \, \Omega$$

The mobilities were found from Table B.3, and the units were not written in the denominator due to lack of space. The units in the denominator match those in the numerator. The answer is negative because the light increases the number of electrons and holes, and therefore the resistance becomes smaller.

8.3   Current Density in a p-n Junction with Light

Before jumping to equations, let us try to qualitatively understand what happens when light is absorbed near a p-n junction. Consider Figure 8.3, where a p-n junction is shown in which a photon is absorbed within the depletion region. Underneath the p-n junction is the electric field profile. The absorbed photon turns into an electron and a hole. The electric field is pointed to the left (negative $\mathcal{E}$). Therefore, the electron, with its negative charge, is pushed to the right by the electric field, contributing to a negative current. The hole, with its positive charge, is pushed to the left, also contributing to a negative current. Thus, when

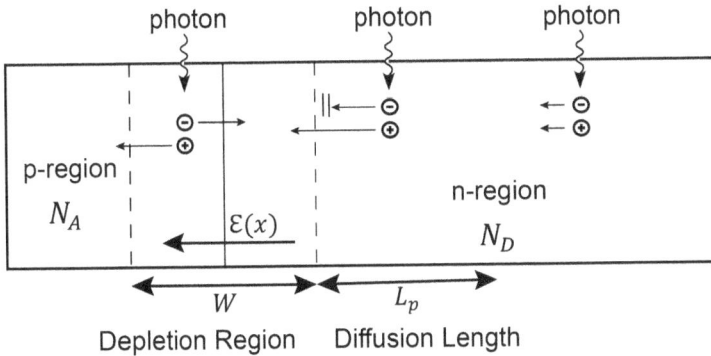

Figure 8.3: In a p-n junction, the behavior of the electrons and holes generated by light depend upon where they are absorbed. If absorbed in the depletion region, the built-in electric field separates the charges, contributing to the optical current. If the photon is absorbed near the depletion region, one carrier (hole in the case shown here) will enter the depletion region, while the other carrier will be prevented from entering due to the built-in electric field. If the photon is absorbed far from the depletion region, it will recombine before reaching the depletion region and does not count for optical current.

light is absorbed in the depletion region of a p-n junction, a negative current density is generated.

Now let's look Figure 8.3, in the region next to the depletion region. Two photons are absorbed in the neutral n-region.

Photon #1: The first photon results in an electron and a hole. There is no electric field, so they are both free to randomly move (diffuse) through the semiconductor. After some time, they both end up at the edge of the depletion region. The electron is blocked by the electric field, and eventually recombines with a hole. The hole is swept to the left by the electric field, contributing to a negative current density, just like holes generated within the depletion region.

Photon #2: The second photon also results in an electron and a hole, but they are formed farther away from the junction than photon #1. There is still no electric field, so they both diffuse through the semiconductor. The electron travels an average distance of one diffusion length, L_n, before recombining, and the hole travels an average distance of one diffusion length, L_p, before recombining.   Since these carriers recombine before reaching the depletion region, they will not contribute to an additional current flow beyond that already calculated for the no-light condition.

These examples illustrate that light will create electrons and holes in the semiconductor.   If the electrons and holes are generated within the depletion region, or within one diffusion length of the depletion region, then the electrons and holes will contribute an additional component to the current density beyond that calculated for a normal (non-illuminated) p-n junction.

To calculate the current due to light (the optical current density J_{op}), let us use the approximation that all photons absorbed within the depletion region, or within one diffusion length of the depletion region, will contribute to current flow.   Then the optical current density is:

$$J_{op} = -q(L_p + L_n + W)G$$
(8.15)

The optical current density is negative because, from Figure 8.3, the optical current density flows in the opposite direction as the diode current density.   The total current density in the p-n junction is:

$$J = J_S\left[\exp\left(\frac{qV_{pn}}{k_BT}\right) - 1\right] - J_{op}$$
(8.16)

The current density for a p-n junction is shown in Figure 8.4a for both an illuminated device and a non-illuminated device.   A very important point is that the current is simply shifted down when light shines on the p-n junction.

Two points are labeled in Figure 8.4: the short circuit current density J_{sc} and the open circuit voltage V_{oc}. The short circuit current density is the current that flows when the applied voltage is zero.   The short circuit current density is the same as the optical current density: $J_{sc} = J_{op}$. The open circuit voltage is the voltage that is measured when the pn junction has zero current flowing.

In reverse bias, the current density is independent of voltage and proportional to the light intensity.   For a voltage larger than the open-

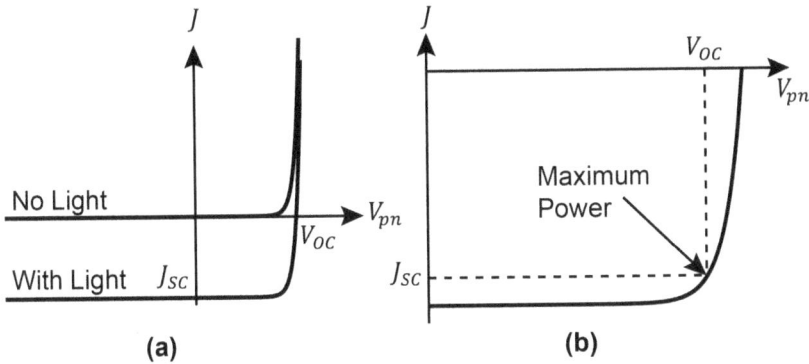

Figure 8.4: (a) Current density in a p-n junction with no light and with light. The curve with light is simply offset from the no-light curve by a constant optical current density, J_{op}. (b) In the fourth quadrant, power is generated. The maximum power is found by maximizing the rectangular area under the curve.

circuit voltage, it is difficult to see the difference in current density between the illuminated and non-illuminated cases because the curves are almost vertical. For small forward bias voltages, below V_{oc}, the current is negative and the voltage is positive, resulting in a negative power dissipation. In this region power is generated and the p-n junction may be used as a photovoltaic cell.

The maximum power that may be obtained from a photovoltaic cell is shown in Figure 8.4b. The maximum power occurs when the rectangular area is maximized. The voltage at which the maximum power occurs depends upon the light intensity. Therefore, it is common for power inverters that connect a solar panel to the electrical grid to use a maximum power point tracking system to ensure that the solar panel delivers the maximum power for the current light intensity level.

For a more accurate current density equation, we should take into account the fact that the light intensity decays as a function of distance, as described by Equation 8.5. Even worse, we should consider the fact that the absorption length differs with wavelength. Addressing these corrections makes for some extremely large equations and we will not tackle them in this book.

8.4   Solar Photovoltaic Cells

Solar photovoltaic cells are p-n junctions that are designed to efficiently generate power from the sun. Solar photovoltaic cells are increasingly attractive as a renewable energy source because their cost has been dropping significantly for many decades now. At the time this book was written, solar power is cost competitive with coal and natural gas in many regions of the United States and parts of the world. Solar combined with battery storage can sometimes compete with fossil fuel-based energy.

While solar photovoltaics are promising, they still suffer from the problem that they only generate power when the sun is shining. This requires that the power grid rely on other power generation technologies and/or energy storage technologies. The proper mix of power generating technologies will be determined in part technical, part economic, part environmental, and part political. This book will not try to address these challenges. Rather, this book will focus on the technical issues related to creating solar photovoltaic cells.

The **figures of merit** that matter most for photovoltaic cells are: **price per watt**, **energy efficiency**, and **lifetime**. While these appear to be straightforward figures, they are complicated by the fact that solar photovoltaics are used in a variety of environmental conditions where extreme temperature variation and solar intensity can influence the numbers. Standard test conditions are at a temperature of 25°C, an incident solar power of 1000 W/m^2 above the atmosphere, and a quantity of air that filters the solar spectrum called the air mass 1.5 spectrum. These parameters are similar to those observed in the United States at a latitude of 48° at noon on the spring and fall equinox.

The **price per watt** is the price of the product divided by the power generated. The price of the product can be the price of the semiconductor, the price of a module, which contains the semiconductor and the fixture to protect the semiconductor from the environment, or the installed price which includes all the permitting, fixturing, and labor for installation. Most often, the price per watt is the module price divided by the peak power output by the module.

The **energy efficiency** is the conversion efficiency of the solar radiation into electrical power. There are standards that are commonly used to rate commercial solar panels. The energy efficiency changes with temperature and it changes with time. Some solar panels, especially thin-film solar panels, degrade significantly in the first few years of their life.

The **lifetime** of a solar panel is not well known. Most commercial solar panels will last for decades, perhaps more than 50 years. The reason is that there are no moving parts and no maintenance is required. The weak point is not the semiconductor material itself, but the module in which it is housed. At present, the inverter that connects the solar panels to the house or grid is the component most likely to fail.

To achieve the greatest amount of power, several items must be optimized:

1. The area must be as large as possible to collect as much light as possible.
2. The p-n junction must cover the entire area because the light must be absorbed within one diffusion length of the junction.
3. The p-n junction must be near the semiconductor surface because that is where the intensity of light is greatest.

Because of these requirements, a solar cell has a geometry similar to that shown in Figure 8.5. The bottom electrode can also act as a

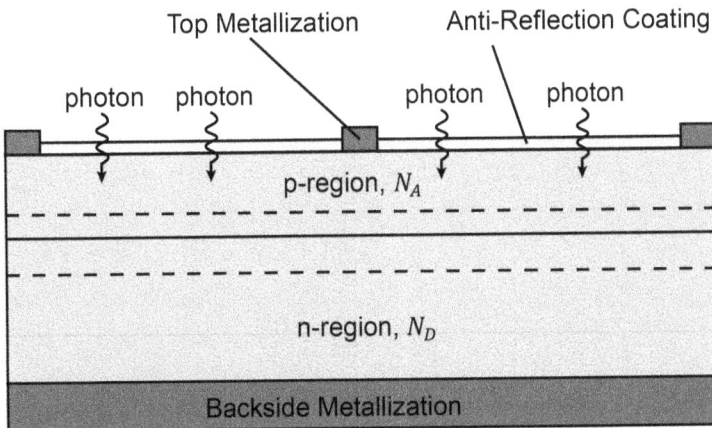

Figure 8.5: Geometry of a typical solar cell.

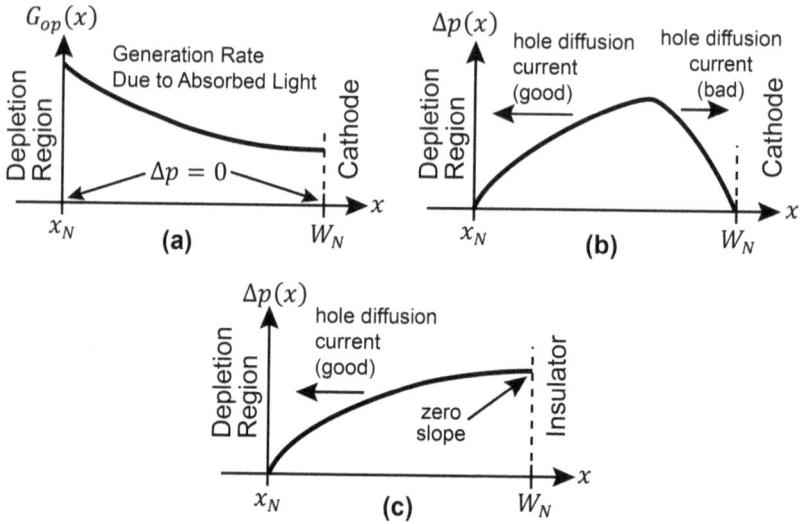

Figure 8.6: (a) The optical generation rate as light is absorbed in the neutral n-region of a solar cell. (b) The minority carrier concentration when the right boundary is a metal. (c) The minority carrier concentration when the right-side boundary is an insulator.

reflective layer to help reflect any light that transmits through the entire semiconductor. The metal on the top is necessary to make a good contact and minimize the series resistance, but its area should be minimized because it reflects light from the p-n junction. An anti-reflection coating is used to get more light into the solar cell because most semiconductor materials have a very high dielectric constant that reflects a significant amount of light. Alternative geometries are also used. For example, the top contact may be eliminated by patterning a p- region on the bottom.

Figure 8.6a shows the generate rate due to light in a solar cell. The light intensity is greatest at the top, and is smaller as the light propagates through the solar cell and the light is absorbed. We can use the continuity equation (Equation 6.28) to calculate the excess hole concentration. But let's analyze this without solving the differential equations. The boundary conditions are $\Delta p = 0$ at the edge of the depletion region, and where the metal contacts the semiconductor. Therefore, we will get a graph for the excess hole concentration that looks similar to Figure 8.6b. It peaks in the center, and goes to zero at either end. Near the depletion region, the holes diffuse to the left. Therefore, there is current to the left, which is the desired direction of current for a solar cell. Near the metal cathode, the holes diffuse to the right. This is in the wrong direction, and reduces the efficiency of the solar cell.

One approach to improving the efficiency of the solar cell is to coat the backside with an insulator. An good insulator will have a much lower recombination rate (ideally zero), the slope of the hole concentration will be zero here. This gives us the graph in Figure 8.6c for the excess carrier concentration. This eliminates the bad hole diffusion current. The only problem is that we still need current flow to the cathode. This can be accomplished by either making very small holes in the insulator (which partially reduces the benefit of the insulator), or by making the insulator so thin that the electrons can tunnel through the insulator.

Photovoltaic Materials

Photovoltaic cells are most commonly made using silicon. Single crystal silicon, which is the material used by the IC industry, achieves the highest silicon efficiency. Polycrystalline silicon is also used because it can reduce the cost of the photovoltaic cell, although the efficiency is not as high. Amorphous silicon is used in a thin film form. The thin film may only be a few μm thick, so a separate low-cost substrate is used for mechanical support. Using only a few μm of silicon helps reduce the cost of amorphous silicon photovoltaic cells, but the efficiency is lower than for single crystal silicon or polycrystalline silicon.

CdTe is another photovoltaic material that is used because it can be produced inexpensively. GaAs, or rather a layered approach of GaAs and similar materials, are used for applications where efficiency is much more important than cost, such as for use in satellites orbiting earth. The cost of GaAs is approximately 1000 times the cost of silicon. Other materials of interest include copper zinc tin sulfide (CZTS) and copper indium gallium selenide (CIGS). Perovskites, which is really a class of materials, are also being pursued because they have shown good efficiencies, but they usually contain lead, presenting environmental concerns. There are also a variety of organic compounds being considered.

The semiconductor used should be chosen with an optimal band gap. If the band gap is too large, many photons with lower energy will not be collected, reducing the efficiency. If the band gap is too small, most of the photons will be absorbed, but most of the energy will be converted to heat because the photon energy is larger than the band gap, as illustrated on the right in Figure 8.1.

The maximum efficiency of a solar cell made with a single material is expressed by the **Shockley-Queisser limit**. This limit is determined using the solar spectrum that strikes the earth, including the effects of atmospheric absorption, which depend upon the location on Earth. It has been determined that the best theoretical efficiency is near 34% for a material with a band gap around 1.1 – 1.3 eV. Silicon, with a band gap of 1.1 eV, has a theoretical maximum efficiency of just over 32%. In practice, the best silicon photovoltaic cells in production (at the time this is written) is over 25%. This shows that there is still room for R&D in both silicon photovoltaics and in R&D on new materials.

Tandem Solar Cells

The Shockley-Queisser limit applies to solar cells made with a single material, with a single band gap. If two (or more) materials are stacked, such that the light first enters a wide band gap material, and then enters a narrow band gap material, the efficiency may be improved. This is called a **tandem solar cell**. The wide band gap material absorbs the light with the higher energy, or shorter wavelengths. The wide band gap material is also transparent to the longer wavelengths of light. Then the narrow band gap material will absorb the longer wavelengths of light. Thus, a photovoltaic cell with multiple band gap materials will absorb a larger spectrum of light, and convert the energy more efficiently, than a single band gap photovoltaic cell. Currently, this approach is far more expensive to produce than single material photovoltaic cells. However, there is a lot of research on this topic because it promises photovoltaic cells with a much larger efficiency.

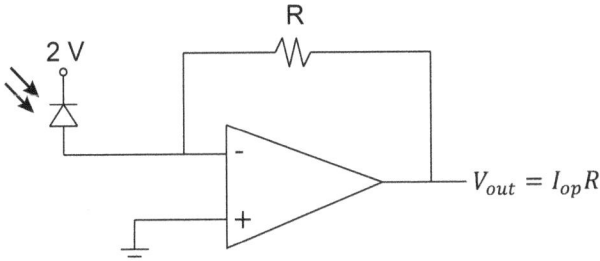

Figure 8.7: Circuit to convert the current from the photodiode into a voltage.

8.5 Photodiodes

Photodiodes are p-n junctions that are operated in reverse bias. Usually, the reverse bias current is very small compared to the optical current. In this case, the current is approximately proportional to the light intensity. The current density is:

$$J = q(L_p + L_n + W)\frac{P_{ph,0}}{E_{ph}L_A} \tag{8.17}$$

Measuring in reverse bias is a great way to measure the light intensity. In forward bias, the voltage and current are both nonlinear with respect to light intensity.

Many applications require a voltage. To convert from a current to a voltage, a transimpedance amplifier may be used. Figure 8.7 shows a simple circuit to bias the p-n junction in reverse bias and convert the current to voltage.

Photodiodes have a variety of applications. Example 8.4 demonstrates how to use four photodiodes as a position sensor.

Example 8.4: An array of photodiodes may be used to determine the position of a laser beam. The following diagram shows four photodiodes arranged in a 2x2 configuration.

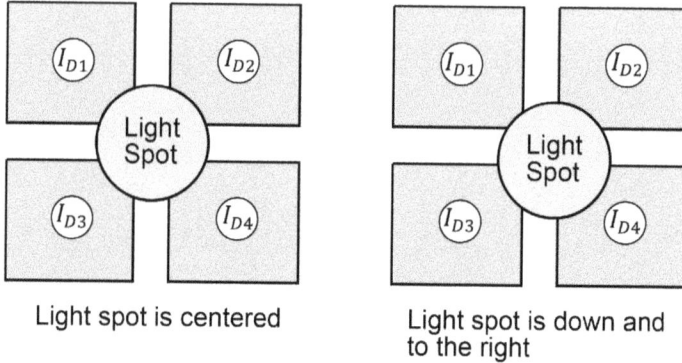

Light spot is centered

Light spot is down and to the right

If the laser spot is exactly center, the current in all four p-n junctions will be identical. If the laser spot is offset, vertically, then the current in the two top photodiodes will be different than the current in the bottom two photodiodes. The y- displacement is proportional the difference between the currents:

$$\Delta y = k_y[(I_{D1} + I_{D2}) - (I_{D3} + I_{D4})] \tag{8.18}$$

Similarly, the x- displacement can be found from:

$$\Delta x = k_x[(I_{D2} + I_{D4}) - (I_{D1} + I_{D3})] \tag{8.19}$$

On an atomic force microscope (AFM), this laser spot detection method may be used to detect the displacement of a cantilever with better than 1 nm accuracy.

8.6 Summary of Light on a Diode

The light on a diode always shifts the current-voltage curve downward (Fig. 8.8). If the diode is operated in quadrant 1 (positive current, positive voltage), then the forward biased diode has a lot of current flowing and the optical current is negligible.

The diode cannot operate in quadrant 2 (positive current, negative voltage) because the light will always move the curve downwards.

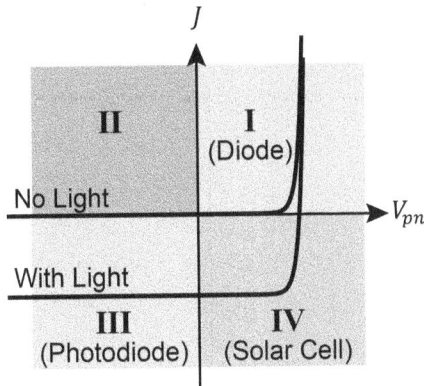

Figure 8.8: Light shifts the curve down. The diode can be operated in quadrant I as a diode, quadrant III as a photodiode, and quadrant IV as a solar cell. The diode cannot be operated in quadrant II.

If the diode is operated in quadrant 3 (negative current, negative voltage), we call the diode a photodiode because the current is proportional to the intensity of light.

If the diode is operated in quadrant 4 (negative current, positive voltage), the diode generates power. This is the only quadrant that generates power, and this is how a solar cell is utilized.

A normal diode, used in electrical circuits, is always packaged in an opaque package to keep light off the semiconductor. This is because the diode will operate in quadrant 1 under forward bias, and quadrant 3 under reverse bias. While the light won't really affect the forward bias, it will significantly affect the reverse bias current (leakage current).

8.7 Summary of Key Concepts

- Light is made up of individual particles called photons.
- A semiconductor can absorb light if the photon energy is greater than the band gap of the semiconductor.
- A semiconductor with a band gap larger than the photon energy is transparent to light.
- When a photon is absorbed, one electron and one hole is generated. This is called an electron-hole pair.

- Light absorption occurs in a semiconductor over distance. The light intensity drops exponentially with distance with a characteristic length called the absorption length.
- The resistance of a semiconductor decreases when light is absorbed because the electron and hole concentrations increase.
- Photons absorbed in the depletion region of a p-n junction, or within one diffusion length of the depletion region, will contribute to an optical current flow through the p-n junction.
- The optical current flow in a p-n junction is in addition to the current that flows due to any applied bias.
- The optical current flow in a p-n junction always shifts the I-V curve downward.
- In reverse bias, the optical current flow is proportional to the light intensity.
- In the fourth quadrant, where the voltage is positive and the current is negative, a negative power is dissipated. This means that power is generated.
- Solar cells are p-n junctions operated such that the current and voltage are in the fourth quadrant of the I-V curve.
- Solar cells made of silicon are currently the most cost effective in terms of cost per watt.
- Multi-junction solar cells are the most efficient solar cells, but are very expensive.

8.8   Problems

1. For light to be absorbed, what relation between the energy of a photon and the semiconductor band gap must be satisfied?

2. A typical lighting level for a room is 200 lux.   1 lux $=$ 1.5×10^{-7} W/cm^2 at 555 nm.   The light shines on a silicon substrate. Assume all the light is monochromatic at 555 nm for this problem. The absorption length in silicon is $L_A = 1.6 \ \mu m$
   a. What is the optical power, P_{ph}, in W/cm^2 ?
   b. What is the optical power at the silicon surface, 1.6 μm below the surface, and 10 μm below the surface?
   c. What is the generation rate, g_{op}, at the silicon surface, 1.6 μm below the surface, and 10 μm below the surface?

3. Light from the sun strikes the Earth with an intensity of 1000 W/m² at the equator.   Let us assume all the light has a wavelength of 555 nm.   Repeat problem #2 for this level of illumination.

> Note: Sunlight has a lot of red and near infrared light, making it desired to make silicon solar cells much thicker than a few micrometers because of the much longer absorption length at those wavelengths.

4. A semiconductor made of silicon is illuminated with light. The semiconductor is uniformly doped with acceptor atoms at a concentration of $N_A = 10^{15}$ cm^{-3}. The light has a wavelength of 600 nm and an absorption length of 100 μm. The lifetime is 100 μs.   You may assume that the electron and hole concentrations are approximately uniform, even with light applied, when solving this problem.
   a. What is the resistivity of the semiconductor with no light?
   b. What is the resistivity of the semiconductor if the light has a power density of 10 mW/cm²?
   c. What is the resistivity of the semiconductor if the light has a power density of 0.01 mW/cm²?
   d. Using a computer, plot the resistivity as a function of the light power density.   Use a log-log plot.   Find an appropriate range of light power density such that you demonstrate the region where light has no effect on the resistivity, and such that you also demonstrate the region where the resistivity changes with light.

5. Consider a GaAs p-n junction with $N_A = 10^{17}$ cm^{-3} and $N_D = 10^{17}$ cm^{-3}. The minority carrier lifetime on the p-side is 1 μs, and the minority carrier lifetime on the n-side is 0.1 μs.
   a. What is the reverse saturation current density, J_S?
   b. What is the current density for $V_{pn} = -3\,V$?
   c. What is the current density for $V_{pn} = 1\,V$?
   d. Plot the current density for an applied voltage ranging from -3 V to 1 V.

e. Light shines on the semiconductor with uniform illumination. The generation rate is 10^{18} cm^{-3}/s. What is J_{op} at 0 V ?

f. Redo the plot for part d. On the same plot, show the current density when light is applied. Make sure both curves are visible.

g. What is the short circuit current density and the open circuit voltage from this level of illumination?

h. What is the maximum amount of power that may be obtained from this level of illumination (W/cm^2) ?

6. Consider a silicon p+-n junction with $N_A = 10^{17}$ cm^{-3} and $N_D = 10^{15}$ cm^{-3}. The minority carrier lifetime on the n-side is 100 µs.

a. What is the reverse saturation current density, J_S?

b. What is the current density for $V_{pn} = -3\,V$?

c. What is the current density for $V_{pn} = 0.6\,V$?

d. Plot the current density for an applied voltage ranging from -3 V to 0.6 V.

e. Light shines on the semiconductor with uniform illumination. The generation rate is 10^{18} cm^{-3}/s. What is J_{op}?

f. Redo the plot for part d. On the same plot, show the current density when light is applied. Make sure both curves are visible.

g. What is the short circuit current and the open circuit voltage from this level of illumination?

h. What is the maximum amount of power that may be obtained from this level of illumination?

7. (Hard) Repeat problem #6, but this time include the effect of a series resistance of $R = 5\,\Omega$. The diode has a cross-sectional area of 1 cm^2.

a. Make a plot that includes light illumination that clearly shows the current in Quadrant IV. Include two curves: one curve that includes the effect of series resistance, and one curve that does not include series resistance. The x-axis should be the voltage across the combined p-n junction and series resistor.

b. What is the difference in the short-circuit current? Explain why your answer makes sense.

   c. What is the difference in the open-circuit voltage? Explain why your answer makes sense.

   d. Explain what happens to the I-V curve near where the output power is maximum. What happened?

   e. Let us quantity the power loss. What is the maximum amount of power that may be obtained from this level of illumination (i) with resistance, and (ii) without resistance? (iii) What is the percent loss in power?

8. The amount of power generated by a solar panel depends upon many factors, including the location on earth, time of year, and amount of cloud cover. The National Renewable Energy Laboratory (NREL), part of the U.S. Department of Energy (DOE), has a nice interface to find this information for most any location in the world. Use the following website to answer these questions for the city in which you live.

pvwatts.nrel.gov

   a. On a monthly basis, what is the smallest amount of solar radiation received in winter? ($kWh / m^2 / day$).

   b. On a monthly basis, what is the largest amount of solar radiation received in summer? ($kWh / m^2 / day$).

9 Light Emitting Diodes and Semiconductor Lasers

In the last chapter, we saw how light can be absorbed in a semiconductor, permitting us to create solar cells and photodiodes. In this chapter, we will learn how light can be emitted by semiconductors, allowing us to create light emitting diodes (LEDs) and semiconductor lasers. LEDs have been used for decades as indicators for electronic devices. More recently, the cost of blue light LEDs has decreased to the point that white light LEDs are competitive with other lighting technologies, such as the incandescent bulb and fluorescent lighting. Semiconductor lasers are used to send optical signals down fiber optic lines used in the backbone of the internet. Semiconductor lasers are also used for optical storage applications, such as the DVD. In the future, optical interconnects using semiconductor lasers may replace copper wires for transmitting information between devices in the home and office. Finally, semiconductor lasers are used to create laser pointers that are used in many presentations and to entertain cats.

9.1 Light Emission in a Semiconductor

To emit light, an electron recombines with a hole. That is, the electron in the conduction band loses energy and occupies an empty state in the valence band. This is a type of carrier recombination. Due to conservation of energy, the light emitted has energy:

$$E_{ph} = E_G \tag{9.1}$$

Not all semiconductors are good emitters of light. We have to satisfy both conservation of energy and conservation of momentum. An electron has momentum equal to:

$$p = \hbar k \tag{9.2}$$

where $\hbar$ is the reduced Plank's constant and k is the wave number of the electron. Thus, momentum and the wave number are related by a

(a) Direct Band Gap
Semiconductor

(b) Indirect Band Gap
Semiconductor

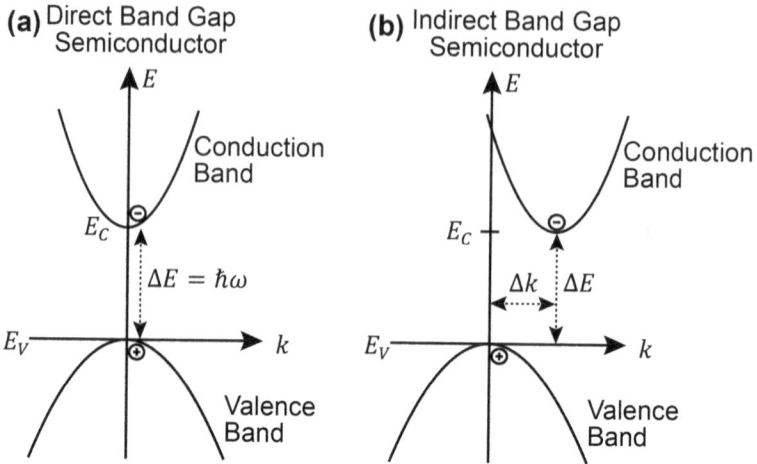

Figure 9.1: (a) Direct band gap semiconductor where the bottom of the conduction band has the same k- value as the top of the valence band.   The direct band gap semiconductor can emit light.   (b) An indirect band gap semiconductor in which the bottom of the conduction band is offset from the top of the valence band.   The indirect band gap semiconductor is very inefficient at emitting light.

constant. Thus far in this book we have shown band diagrams in which the y- axis is energy and the x- axis is position. To better understand semiconductors, we need to show a new type of band diagram, called an E-k diagram, which relates the energy (y-axis) to the (scaled) momentum (x-axis, or k-axis).

Figure 9.1 shows two E-k diagrams for two different semiconductors. Let's start by looking at Figure 9.1a.   The electrons have zero momentum ($k = 0$) when the energy of the electron is equal to E_C, the bottom of the conduction band. If the electron has a higher energy, then the electron gains kinetic energy and has momentum ($k \neq 0$). Most of the electrons are at their lowest energy, and hence most of the electrons have $k \approx 0$. The holes behave similarly, except an increase in energy for a hole moves the hole down on the energy scale for electrons. The lowest energy for a hole is at the top of the valence band, E_V, and most holes have $k \approx 0$. Light has momentum, but the momentum is about 1/1000 of the momentum of electrons.   Therefore, to conserve momentum, the electron has to drop nearly vertically when giving off light. Thus, the electrons may <u>directly</u> recombine with the holes. A semiconductor with the E-k diagram shown in Figure 9.1a is called a **direct band gap semiconductor**.

An interesting property of some semiconductors is that the momentum of electrons at the bottom of the conduction band does not need to be zero. Figure 9.1b shows such an example. This is called an **indirect band gap semiconductor**. For the electron to move from the conduction band to the valence band, the electron has to change momentum as well as energy. The required change in momentum equal to $\hbar\Delta k$. Since the momentum of light is too small to provide this momentum, the momentum has to be provided by vibrations in the crystal. Thus, for the indirect band gap semiconductor to emit light, an electron must be in the same place as the hole, and the atom at this location has to be vibrating with just the correct amount of momentum. The odds of all three occurring in the same place is extremely small. Thus, this kind of semiconductor emits light very poorly.

> In summary, all semiconductors absorb light. Direct band gap semiconductors are good emitters of light. Indirect band gap semiconductors are very poor emitters of light.

GaAs is a direct band gap semiconductor, and GaAs is commonly used to create red LEDs. The band gap of GaAs may be changed by alloying it with various other compounds, permitting other colors to be emitted near red. GaN is a direct band gap semiconductor that gives off blue light.

Silicon, Germanium, and SiC are indirect band gap semiconductors and give off very little light. If you take a silicon diode and run current through it, you can measure the light emitted using a photomultiplier tube. This level of light is not useful for practical applications.

Example 9.1: GaAs is a direct band gap semiconductor. What wavelength of light is emitted?

Answer: The energy of the emitted photon is the same as the energy of band gap. Solving Equation 8.2 for the wavelength we find:

$$\lambda = \frac{1.24 \ (\text{eV·}\mu\text{m})}{E_{ph}} = \frac{1.24 \ (\text{eV·}\mu\text{m})}{E_g} = \frac{1.24 \ (\text{eV·}\mu\text{m})}{1.42 \ \text{eV}} = 0.873 \ \mu\text{m}$$

This is in the infrared range. To make red LEDs, it is common to alloy GaAs with another material to increase the band gap.

9.2 Light Emitting Diodes (LEDs)

The emission of light requires the recombination of an electron and a hole. Thus, we need to have a large number of electrons and a large numbers of hole in the same location to give off a lot of light. First consider an n-type semiconductor: It has a lot of electrons, but very few holes. Now consider a p-type semiconductor. It has a lot of holes, but very few electrons. And an undoped semiconductor has very few electrons and very few holes. So what do we do?

To get a lot of electrons and holes, it is necessary to have a large excess concentration of carriers. A p-n junction, or diode, has a large excess electron concentration and a large excess hole concentration when it is forward biased. Thus, a diode that is made from a direct band gap semiconductor is a light-emitting diode, or LED. An LED gives off a single wavelength of light equal to the band gap of the semiconductor.

In Chapter 7, we saw that under forward bias a p-n junction has a large excess hole concentration in the n- region, and a large excess electron in the p- region. In the n- region, the excess holes can recombine with the large electron concentration, giving off light. In the p- region, the excess electrons can recombine with the large hole concentration, giving off light. What we neglected in Chapter 7 was an analysis of what is occurring in the depletion region. Figure 9.2 shows the electron and hole concentration in a p-n junction, including the majority carriers. To be able to see the minority carriers on a single plot requires a log scale for the y-axis. In the depletion region are dotted lines connecting the

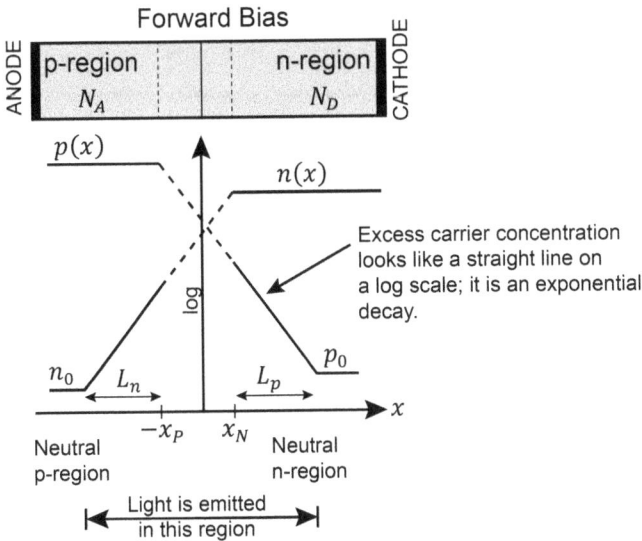

Figure 9.2: Light is emitted from a LED in the depletion region and approximately within one diffusion length of the depletion region.

majority carrier concentration with the excess minority carrier concentration that we calculated in Chapter 7. We are not going to solve for the electron and hole concentration in the depletion region because it requires solving for both the drift current and the diffusion current. The drift current is present because of the electric field. The diffusion current is present because of the large change in concentration over a small distance. This is difficult to solve and is normally solved numerically. We will only show an appropriate solution by connecting the known values in the neutral n- and p- regions. Looking at the figure, we notice two things: (1) the depletion region still has a lot of electrons and holes, despite its name, and (2) there are a lot of electrons and holes that can recombine and give off light. In summary, light is given off in an LED in the depletion region and approximately within one diffusion length of the depletion region.

An LED is a diode. It has the same characteristics: small forward voltage drop, small reverse saturation current, and a breakdown voltage. In fact, the I-V curve looks identical to a normal diode. People often compare LEDs to silicon diodes. The larger band gap necessary to emit visible light means that the reverse saturation current of an LED is significantly smaller than for a silicon diode. Similarly, the forward current of an LED is much smaller at the same voltage, and thus the forward voltage drop to attain the same current density is much larger. It is often said that the turn on voltage of a silicon diode is 0.7 V and the turn on voltage of a LED is 1.4 V. GaN, used to make a blue LED, will have a significantly larger turn on voltage.

Example 9.2: It is desired to create a green LED. What should the band gap be, and what material could we use?

Answer: The energy of the emitted photon is the same as the energy of band gap. Solving Equation 8.2 for the wavelength we find:

$$E_g = E_{ph} = \frac{1.24 \ (\text{eV·μm})}{\lambda} = \frac{1.24 \ (\text{eV·μm})}{0.5 \ \text{μm}} = 2.48 \ \text{eV}$$

The list of semiconductors in Appendix A does not have a material with a band gap near 2.5 eV, but a literature search yields that GaP doped with nitrogen is an appropriate material to make a green LED.

Example 9.3: An infrared LED is made using GaAs. The n-region is doped with $N_D = 10^{18} \ \text{cm}^{-3}$ donor atoms, and the p-region is doped with $N_A = 10^{19} \ \text{cm}^{-3}$. A forward bias of 1 V is applied to the LED. The lifetime in both regions is 0.1 μs. What are (a) the excess carrier concentrations at the edge of the depletion region, and (b) the recombination rate at the edge of the depletion region? (c) If the forward bias is increased to 1.1 V, how much brighter would the LED be?

Answer: (a) The excess carrier concentrations are found from Equations 7.23 and 7.27:

$$\Delta p(x = x_N) = \frac{n_i^2}{N_D}\left[\exp\left(\frac{qV_{pn}}{k_B T}\right) - 1\right]$$

$$= \frac{(1.77 \times 10^6 \ \text{cm}^{-3})^2}{10^{18} \ \text{cm}^{-3}}\left[\exp\left(\frac{1 \ \text{V}}{0.0259 \ \text{V}}\right) - 1\right]$$

$$= 1.77 \times 10^{11} \ \text{cm}^{-3}$$

$$\Delta n(x = -x_P) = \frac{n_i^2}{N_A}\left[\exp\left(\frac{qV_{pn}}{k_BT}\right) - 1\right]$$

$$= \frac{(1.77 \times 10^6 \text{ cm}^{-3})^2}{10^{19} \text{ cm}^{-3}}\left[\exp\left(\frac{1\text{ V}}{0.0259\text{ V}}\right) - 1\right]$$

$$= 1.77 \times 10^{10} \text{ cm}^{-3}$$

(b) The recombination rates are found from Equations 6.23 and 6.24:

$$R(x = x_N) = \frac{\Delta p(x = x_N)}{\tau} = \frac{1.77 \times 10^{11} \text{ cm}^{-3}}{10^{-7}\text{ s}}$$

$$= 1.77 \times 10^{18} \text{ cm}^{-3}\text{ s}^{-1}$$

$$R(x = -x_P) = \frac{\Delta n(x = -x_P)}{\tau} = \frac{1.77 \times 10^{10} \text{ cm}^{-3}}{10^{-7}\text{ s}}$$

$$= 1.77 \times 10^{17} \text{ cm}^{-3}\text{ s}^{-1}$$

(c) The brightness of an LED is determined by the number of photons emitted per second. In general, this is found by integrating the recombination rate over the volume of the LED. We can simplify this calculation if we notice that the brightness is proportional to $\exp\left(\frac{qV_{pn}}{k_BT}\right)$. Therefore, the ratio of the LED brightness at 1.1 V to the LED brightness at 1.0 V is:

$$\frac{\exp\left(\frac{qV_{pn2}}{k_BT}\right)}{\exp\left(\frac{qV_{pn1}}{k_BT}\right)} = \frac{\exp\left(\frac{1.1\text{ V}}{0.0259\text{ V}}\right)}{\exp\left(\frac{1\text{ V}}{0.0259\text{ V}}\right)} = 47.5$$

In summary, increasing the voltage by 0.1 V will increase the brightness by 47.5 ×.

9.3   White Light LEDs

LEDs only give off a single color. This is a problem for general illumination applications, such as to illuminate a room. To overcome this problem, white light LEDs are made using the combination of a blue LED and phosphors. A blue LED made of GaN is used to generate the blue color of the visible light spectrum, and a coating of phosphors are used to absorb the blue light and re-emit a variety of lower energy photons that make up the other colors of the visible light spectrum. It is possible to absorb blue light and emit green light, with the difference in energy being lost as heat. It is impossible to absorb a single green photon and turn it into a blue photon because this process increases the photon energy and violates the law of conservation of energy.

9.4   Semiconductor Lasers

A laser can be made using a semiconductor. A laser consists of several components: (1) an optical gain medium, and (2) an optical cavity. There are two key parameters that must be satisfied for lasing to occur:

1. Population inversion: The number of electrons in a higher energy state must be greater than the number of electrons in a lower energy state. In a semiconductor laser, this is achieved by making the concentration of electrons at the bottom of the conduction band greater than the concentration of electrons at the top of the valence band.

2. Optical gain greater than 1: Using mirrors, an optical cavity is created in which the optical gain is greater than 1. That is, the light bounces back and forth between the mirrors, increasing in optical intensity. A small amount of light is let out at the end to create the laser beam, but the amount of light let out has to be less than the amount of optical gain in a single pass.

Direct band gap semiconductors have been used to make semiconductor lasers. GaAs is a very popular material to use. The dielectric constant of GaAs is so high that the amount of light reflected at the GaAs-air interface can be sufficient to make a laser.

To get population inversion, a p-n junction must be forward biased with a lot of current. Population inversion requires that we operate in the so-called high-level injection region, where the excess carrier concentration is greater than the doping concentration. Therefore, the equations for the p-n junction in Chapter 7 are no longer accurate, as we assumed low-level injection as the equations were derived. Therefore, this section will only give a brief overview of semiconductor lasers.

A typical geometry for a semiconductor laser is shown in Figure 9.3. The depletion region has a very high electron concentration and a very high hole concentration, and population inversion occurs in the

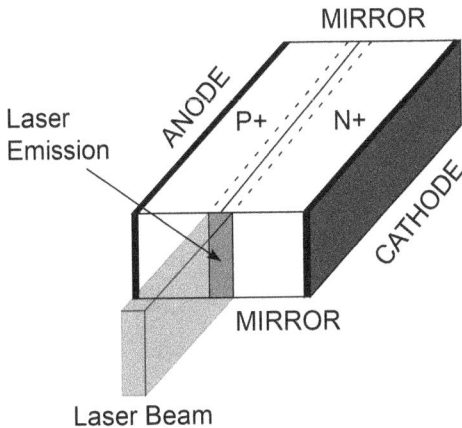

Figure 9.3: Schematic shows a semiconductor laser formed by a p-n junction and using mirrors formed by cleaving the semiconductor.

depletion region. The light passes back and forth through the long direction of the diode, being reflected by mirrors on each side. The mirrors are not perfect so that a laser beam can be emitted through the mirror. Population inversion requires a lot of current, and a semiconductor laser made using this simple geometry is impractical and cannot operate at room temperature.

There are several features that are commonly incorporated to make a better semiconductor laser. First, one may add Bragg reflection mirrors to carefully control the mirror reflectivity. Second, the electron and hole concentrations are not constant in the depletion region, and the electron concentration and hole concentration do not peak at the same location, as shown in Figure 9.2 for an LED. A heterojunction may be used that confines the electrons and holes in a small region with a smaller band gap, as shown in Figure 9.4. The high electron concentration on the n-side spills into the narrower band gap region, where the electrons get trapped. Similarly, the high hole concentration on the p- side spills into the same narrower band gap region, where the holes get trapped. The concentrations of both the electrons and holes are very high in the same region, improving the laser efficiency. Third, since the excess electrons and holes are generated next to the narrow band gap region, the material for the narrow band gap may be lightly doped to decrease the number

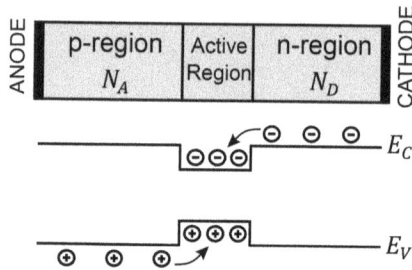

Figure 9.4: Semiconductor laser that uses a narrower band gap region for the active optical cavity. This region traps the electrons and holes, increasing the laser efficiency. The laser beam is emitted perpendicular to the paper.

of defects. Fourth, this configuration also permits simpler control of the optical cavity in the transverse directions because a material may be chosen with a larger dielectric constant in the optical cavity, similar to the way an optical fiber confines light.

Semiconductor lasers find many applications. They are used in CD and DVD players. They are used for laser pointers. And they are used for transmitting information over fiber optics. Semiconductor lasers are replacing conventional lasers used in other applications because they are generally less expensive, more efficient, smaller, and much more reliable than conventional lasers. The only limitation is the development of a direct band gap semiconductor material that emits light at the appropriate wavelength.

9.5   Summary of Key Concepts

- Light can only be emitted from a semiconductor when an electron and hole recombine **AND** the electron and hole had the same momentum prior to recombination.
- Semiconductors in which the electrons and holes have the same momentum are called direct band gap semiconductors.
- Semiconductors in which electrons and holes have different momentum are called indirect band gap semiconductors.
- The energy of a photon emitted from a semiconductor is the same as the band gap energy.

- An LED is made from a p-n junction formed in a direct band gap semiconductor. When the diode is forward biased, the excess electrons and excess holes recombine and give off light.
- White light LEDs are really blue light LEDs that include a phosphorescent material to convert some of the blue light to the other colors of the visible light spectrum.
- Lasers require a population inversion to operate. In a semiconductor laser, this is achieved by biased a p-n junction such that the electron concentration in the conduction band is greater than the electron concentration at the top of the valence band.
- Semiconductor lasers are devices that incorporate an optical gain medium, similar to an LED, and an optical cavity formed using mirrors.
- Semiconductor lasers typically utilize heterojunctions that trap electrons and holes in the same place by using a narrow band gap semiconductor region for the active region.

9.6   Problems

1. There are two conservation laws that must be satisfied for a semiconductor to give off light. What are the two conservation laws?

2. Draw the band diagram of (a) a direct band gap semiconductor, and (b) an indirect band gap semiconductor.

3. Which of the following semiconductors have a direct band gap?
   a. Silicon
   b. Germanium
   c. GaAs
   d. GaN
   e. SiC

4. When an electron and a hole recombine, how many photons are given off?

5. An LED is made using GaN. What wavelength of light is given off? What color does this correspond to?

6. Consider silicon.

   a. If silicon were to emit light, what would the wavelength be?

   b. What is the wavenumber of the light? $k = 2\pi/\lambda$

   c. What is the momentum of the light? $p = \hbar k$

   d. What is the momentum of an electron at the bottom of the conduction band, using $k = 0.8\,\pi/a$, and $a = 0543$ nm.

   e. How do the momentum of the electron and light compare? Does this explain why silicon does not emit light?

10   Economics of Semiconductor Device Manufacturing

Everyone knows that computers, laptops, tablet computers, cell phones, game consoles, and so on, become better and cheaper every year. This occurs because engineers make transistors smaller, faster, and less expensive. The combination of significantly improved performance at a significantly lower cost is rare to find in most industries, and yet it has happened for decades in the semiconductor industry.

This chapter will start with an overview of how semiconductor devices are manufactured, followed by an analysis of how to calculate the die yield on a wafer. Then the economics of producing semiconductor devices will be discussed. Finally, we will look at the capital costs for semiconductor manufacturing.

A few definitions are in order. A **wafer** is a thin slice of a semiconductor on which the semiconductor devices are fabricated. Each device is called a **die** while it is either on a wafer or after being separated from the other die on a wafer. Each wafer typically contains hundreds or thousands of die. After the device is put in a package, it is called a **chip** or other identifying name such as LED or laser. For simplicity, this chapter will refer to the completed, packaged device as a **chip**. If you look at a device and are looking at the semiconductor material, then it is a die. If you see the plastic or ceramic package covering the device, then it is a chip.

Figure 10.1 show an example of a photomask, wafer, die, and a die in an open package. A wafer typically contains hundreds or thousands of die on it. The wafer is cut into individual pieces, called die. The die is too fragile for handling, so the die is enclosed in a package that provides wires for soldering as well as to protect the die. The wires on the die are too small to solder a wire to. You can see the pins on the package, but you cannot see the tiny aluminum squares on the silicon die.

Figure 10.1: A photomask, a (small) silicon wafer with many die on it, an individual die from a wafer, and a die in an open package.

10.1 Semiconductor Device Manufacturing

Semiconductor devices are manufactured using a slice of semiconductor, called a wafer, which is up to 450 mm in diameter. Silicon wafers of up to 300 mm diameter are routinely used, and 450 mm diameter silicon wafers are used in research labs. SiC and GaAs wafers may be up to 150 mm in diameter. Other semiconductors are typically smaller in diameter, as they do not have the manufacturing volume to justify the capital expenses and research budget required to make large diameter wafers with sufficiently low defect densities. Typically, multiple devices are manufactured on a single wafer.

Figure 10.2 shows a schematic of the process to manufacture a semiconductor device. The starting material is a semiconductor wafer. Three types of processes are performed on a wafer: (1) a deposition step is performed to deposit material everywhere on the substrate, (2) a photolithography step is used to apply a pattern on the substrate using a coating that protects the regions where we want to keep the deposited material, and (3) an etching step is performed to remove any unwanted material. These steps are repeated as many times as necessary to process the wafer to its final state.

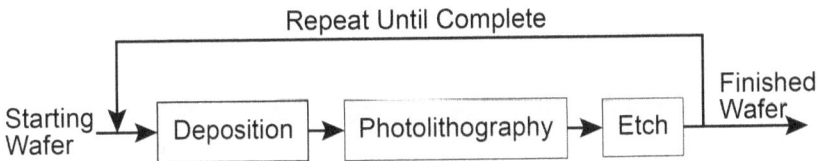

Figure 10.2: Typical processing steps that are used to create a finished wafer.

These steps are performed uniformly on the entire wafer.

Deposition is the process of adding material to the top of the substrate, or even within the substrate. The deposition step will coat the entire wafer.

> **Deposition Example #1**: a layer of aluminum may be deposited on the entire substrate.

> **Deposition Example #2**: SiO_2 may be chemically grown on silicon by reacting the silicon substrate with O_2 gas at high temperature. $Si + O_2 \rightarrow SiO_2$.

Photolithography is the procedure of using light to expose a light-sensitive polymer on the substrate. Light is incident on a **photomask.** An example of a photomask is shown in Figure 10.1. Light passes through the transparent glass. The chromium image on the photomask is transferred to the substrate to define a pattern. The following steps are performed:

> **Step #1**: Coat the wafer with a thin layer of **photoresist** – a light sensitive polymer.

> **Step #2**: Using an imaging system, expose the photoresist to the light source through a photomask. The portion of the photoresist exposed to light will be chemically modified.

> **Step #3**: Develop the photoresist. Insert the wafer into a developer that removes the chemically modified photoresist that was exposed to light, and leaves intact the photoresist that was not exposed to light.

Etching is the process of removing material from the substrate. This is generally some form of chemical process. It is often possible to enable the etchant to remove the desired material while not attacking other materials on the wafer.

> **Etch Example #1**: SiO_2 may be removed from a silicon substrate using HF acid; the HF etches the SiO_2, but does not etch the silicon.

> **Etch Example #2**: A wafer contains an aluminum layer with photoresist patterned on it. A plasma containing chlorine gas is used to chemically etch the aluminum. The chlorine gas does not attack the photoresist. The aluminum coated by photoresist is not removed, as the photoresist protects those regions of the wafer. The aluminum that is not covered by photoresist is etched from the wafer.

Example 10.1: The process described above and illustrated in Figure 10.2 is an example of batch fabrication. When wires are attached to a transistor, ALL the wires are attached at once through the three steps of deposition of aluminum, creating the desired pattern of aluminum using photolithography, and etching the aluminum from the regions that are not part of a wire. Let us contrast this batch process to a serial process in which a robot attaches each wire individually.

Consider a super robot capable of attaching 1000 wires per second. It will be used to attach three wires to every MOSFET on an integrated circuit for the source, drain, and gate. And consider a modern integrated circuit with 1 Billion transistors on it. How long would it take to attach all the wires?

Answer: There are 3×10^9 wires to be attached, at a rate of 1000 wires per second. This works out to 3×10^6 seconds, or 35 days to attach the wires. Of course, no such super robot currently exists, and a more reasonable number might be 10 wires per second, making the time to attach the wires nearly 10 years!

If wires were attached sequentially, the cost of an integrated circuit would be thousands or even millions of dollars. And yet, I can purchase an integrated circuit with billions of transistors for less than $1 USD. This is possible because of batch fabrication.

Example 10.2: Engineers are clever. Sometimes the sequence deposit, lithography, etch described in Figure 10.2 is not followed. For example, consider how Boron dopant atoms are inserted in a silicon wafer to create a p-region. First, photolithography is performed to apply a patterned layer of photoresist, covering the portions of the semiconductor that should not have the dopant atoms. Boron dopant atoms are introduced into the top portion of the wafer using ion implantation – a technique to force atoms into the surface of a solid using a particle accelerator. Wherever there is photoresist, the Boron atoms go into the photoresist. Wherever there isn't photoresist, the Boron atoms go into the silicon. Thus, once the photoresist is removed, we have a silicon wafer with p-regions only where we want. This process did not require an etch step; just photolithography and deposition.

The wafer is processed using the sequence of deposition, lithography, and etch until the desired devices are fabricated on the wafer. This is the end of the batch fabrication techniques. From here, the processing steps are performed on individual devices, called die. The final steps are shown in Figure 10.3. Each device is individually tested using a preliminary testing procedure to determine the bad devices so that time and money isn't spent on bad die. This level of testing is called **wafer-level testing**.

The wafer is separated into individual devices using a dicing saw. The dicing saw is similar to a circular saw used to cut wood, except it uses a diamond blade. Each die may now be picked up individually for further processing. The good die are kept, and the bad die are discarded or analyzed in a failure-analysis lab.

Each good die is inserted into an IC package wireframe, and wires are attached between the IC package wireframe and the die using a special machine called a wire-bonder. The wire-bonder attaches the thin wires using ultrasonic welding, and each wire has to be attached individually. An automatic wire bonder that attaches multiple wires per second is really an amazing thing to see.

Finished Wafer

```
   ┌──────────────┐
   │ Wafer Level  │
   │    Test      │
   └──────────────┘
   ┌──────────────┐
   │  Dice Wafer  │
   └──────────────┘
       ┆ - - - -▶  Discard
       ┆           Bad Die
   ┌──────────────┐
   │   Package    │
   │     Die      │
   └──────────────┘
   ┌──────────────┐
   │  Final Test  │
   └──────────────┘
```

Finished!

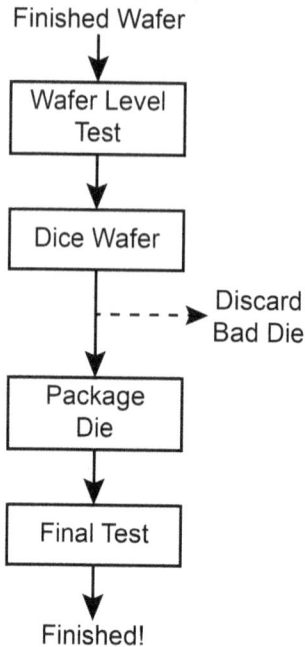

Figure 10.3: After processing the wafer, each die on the wafer is tested. The wafer level test is a preliminary test designed to rapidly identify bad die, and need not be an exhaustive test. Next, the wafer is diced. The bad die are discarded to avoid the cost of packaging the bad die. Then the good die are packaged, and a final test is performed. The final test verifies that every transistor functions, and that the final device is ready to be sold.

Finally, the plastic coating around the IC is formed using plastic injection molding. The final device is tested again to ensure that it performs as advertised.

A key point to keep in mind is:

> The cost to process a wafer is largely the same, regardless of the wafer size. There is only a small increase in cost as the wafer size increases.

10.2 Cost Analysis

From a manufacturing point of view, cost is a key determining factor for how the wafers are processed. The cost of a piece of equipment is less important than the cost per wafer processed. For example, a machine that costs $1M and takes 10 minutes to process a wafer, or a machine that costs $10M and takes 1 minute to process a wafer, has the same cost per wafer.

The number of die on a wafer determines the actual die cost. If the die can be made smaller, using smaller transistors, then there will be more die on a wafer. For a fixed manufacturing cost per wafer, the cost per die is therefore smaller.

The total cost of a chip includes the following items:

Fixed Costs ($Fixed): The fixed costs include the engineering costs to carry out the chip design and the cost to make the photomasks. The photomask costs are a one-time cost, independent of the number of wafers processed. The cost of photomasks has been steadily increasing, making it hard to amortize the fixed costs when only a small number of devices are manufactured.

Wafer Cost ($Wafer): The cost of a wafer is the total cost of all the etching, deposition, and lithography steps required to fabricate the wafer.

Package Cost ($Package): The die has to be encased in a plastic or ceramic package to protect the die and attach leads for easy soldering. Plastic packages are the least expensive. Ceramic packages are more expensive, but provide better environmental protection and can transfer more heat. Packages come in many forms, including DIP packages and surface mount packages.

Test Cost ($Test): At the wafer level, all of the die are tested in order to quickly sort out the bad die from the good die. After packaging, every transistor and every wire will be tested to make sure they function properly. Timing information will also be measured. This test is obviously quite exhaustive, and much engineering effort goes into optimizing the test procedures. Test engineering is a field of its own.

Yield, or Die Yield (Y): The die yield is the percent of good die on a wafer. If a dust particle lands on the wafer during fabrication, it can cause a manufacturing defect that makes a circuit fail. No manufacturer has a 100% yield. Some processes can approach 100% yield, but manufacturers will often start selling chips when the yield is as low as 10%, with the intention of improving the yield as they gain manufacturing experience by selling a lot of chips.

Die Cost ($Die): This is the cost of an individual die on a semiconductor wafer. This cost does not include the one-time costs, packaging costs, or testing costs. The die cost is the wafer cost divided by the number of good die per wafer. That is, the die cost is:

$$\$Die = \frac{\$Wafer}{\#Good\ Die} = \frac{\$Wafer}{\frac{A_{Wafer}}{A_{Die}} \times Y} \tag{10.1}$$

where A_{Die} is the area of a single die on a wafer and A_{Wafer} is the total area of a wafer. The denominator is simply the number of good die on a wafer.

Chip Cost ($Chip): This is the cost of the final chip, including the wafer cost, package cost, and testing cost. Sometimes the chip cost will include the one-time costs, but this requires knowledge of the total number of chips to be produced. If the volume is not known, then the chip costs may not include the one-time costs.

Putting all these costs together, the chip cost is:

$$\$Chip = \frac{\$Fixed}{N_{chips}} + \$Die + \$Package + \$Test \tag{10.2}$$

where N_{chips} is the total number of chips to be produced.

For some ICs, the package and test costs are greater than the die cost. This is particularly true for ICs that cost $0.10 each. For other ICs, the die cost dominates. This is particularly true for high performance CPUs, GPUs, and memory.

10.3 Die Yield

The die yield, or simply yield, is a very important parameter. The die yield has a larger impact on the chip cost than a quick glance at the cost equations may indicate. Obtaining real-world yield information is difficult because no company publishes their yields.

The importance of the yield is evidenced by the fact that manufacturers use cleanrooms to fabricate their devices. The cleanroom provides a clean environment, largely free of dust particulates, in order to reduce defects and improve yield. But a modern cleanroom facility costs over $1 Billion just for the building, not including any of the equipment inside. If the cleanroom was not necessary to improve their yield, then manufacturers would not spend such sums of money.

The simplest model for calculating yield is the **Poisson model**. Each defect is assumed to be independent of all other defects. There is a manufacturing defect density, D, that represents the defect rate per area. Typical units are # defects / cm^2. The chance of a defect occurring on a die depends on the area of the die. The larger the die, the greater the odds of a defect. There is a chance that a single die has two (or more) defects, but this would only result in one bad die. From probability theory, the statistics that describe this situation is the Poisson model, and the equation for the yield is:

$$Y = \exp(-DA_{\text{Die}}) \qquad (10.3)$$

where A_{Die} is the area of a single die. Notice that the yield does not depend on the size of the wafer.

Example 10.3: Consider the case where the defect density is 1 defect per cm^2, and the area of an individual die is 1 cm^2. The yield is found to be 37%. That is, on average, there is one defect per die, but some die have more than one defect, and 37 % of the die have zero defects. 37 % of the die on a wafer are good die, and 63 % of the die on a wafer are bad die.

If we consider the fact that the defects are not independent, a more accurate method for calculating yield may be obtained. A good example of this is if you dropped a flower pot on the floor. There would be dirt scattered on the floor, but the dirt is not uniformly dispersed. A greater concentration of dirt will be found underneath the area where the flower pot fell, with less dirt farther away. From a semiconductor manufacturing point of view, when one defect occurs, it is likely that a second defect occurs nearby. Another common method for calculating yield is the **binomial distribution**, where a parameter α is fit to the experimental data to represent the clustering of defects. Typical numbers for a CMOS process have α between 3 and 4. The yield equation is:

$$Y = \left(1 + \frac{DA_{\text{Die}}}{\alpha}\right)^{-\alpha} \qquad (10.5)$$

Example 10.4: Consider the case where the defect density is 1 defect per cm^2, $\alpha = 3$, and the area of an individual die is 1 cm^2. In example 10.3, assuming no clustering of defects, the yield was calculated to be 37%. With clustering, the yield is calculated to be 42 %. That is, clustering results in more good die because when defects occur, the defects are more likely to be concentrated within a single die.

10.4 Smaller Transistors Are Less Expensive

The semiconductor industry has always strived to reduce the size of a transistor. The smaller the transistor, the more transistors that may put in a given area, and the smaller the overall chip size. In this section we will see that smaller transistors are much, much less expensive than larger transistors.

Example 10.5. Consider a manufacturer who makes ICs on a wafer at a cost of \$4,000 per wafer. The wafer diameter is 300 mm, and each chip requires 1 cm^2 of area. The defect density is 0.3 defects / cm^2. What is the yield using Poisson statistics, the number of good chips per wafer, and the cost per chip?

Now consider that the manufacturer adopts the next generation lithography node, reducing the gate length by 0.7. The wafer cost increases 20%, but the defect density remains the same. What is the yield, number of good chips per wafer, and cost per chip?

The following table summarizes the two processes:

	Original Process	**New Process**
Wafer Cost	\$4,000	\$4,800
Wafer Diameter, A_{Wafer}	300 mm	300 mm
Die Size, A_{Die}	1 cm^2	0.5 cm^2
Defect Density, D	0.3 defects/cm^2	0.3 defects/cm^2
Yield, $Y = \exp(-DA)$	74 %	86 %

# Good chips per wafer, $\dfrac{A_{Wafer}}{A_{Die}}Y$	522	1214
Cost per chip, $\dfrac{\$Wafer}{\text{\# Good chips}}$	$7.66	$3.95

Using the new process, the cost per chip is nearly half the price of the old process, despite a 20% increase in wafer processing cost. This is an amazing result, and it provides a very strong reason to push very hard to make the smallest transistors possible.

Can you imagine working for a company that is one generation behind your competitor, with a manufacturing cost nearly twice that of your competitor? Your competitor can easily price their product below your product manufacturing cost, and they will still make a healthy profit. Your company will be forced to match the price and lose money on everything you sell. A company that is one generation behind the competition will go bankrupt.

10.5 MOSFET Scaling

Historically, the IC industry introduced a new generation of products every 2 years whose dimensions are 0.7x smaller, resulting in a scaling of 0.5x smaller area. This is known as Moore's Law, and accurately described the rate of progress in transistor scaling for many decades. The following list shows some typical minimum feature sizes used in manufacturing over the years:

90 nm
65 nm
45 nm
32 nm
22 nm

The minimum feature size that photolithography can produce is known as a manufacturing node. Due to changes in how MOSFETs are manufactured, the node number no longer refers to a feature size. In recent years, the rate of advancement has slowed, but MOSFETs are still getting smaller because of the strong financial incentive to do so.

As transistors are made smaller, the trend is for all dimensions to shrink, including the gate insulator thickness and the thickness of the various depletion layers. The operating voltage tends to decrease, and the threshold voltage tends to decrease as much as possible in order to maintain a large $V_{GS} - V_T$. Finally, the overall capacitance tends to decrease as transistors become smaller.

The net result is that as transistors become smaller, power consumption per transistor is reduced significantly and the time to turn a gate on or off is reduced. That is, **we get faster circuits that run on less power and at lower cost by making transistors smaller**.

10.6 Summary of Key Concepts

- Semiconductor devices are fabricated on a slice of semiconductor material called a wafer.
- Multiple semiconductor devices are fabricated on a single wafer. The more devices per wafer, the lower the cost per device.
- The processing of a wafer consists of three basic steps that are repeated many times: photolithography, etching, and deposition.
- The cost to process a wafer is independent of the number of devices on a wafer.
- A preliminary test is performed on all the devices at the wafer level after fabrication. This permits the manufacturer from spending the cost to package a bad device.
- Final test is performed after the device is fully packaged.
- The number of good die on a wafer depends upon the manufacturing maturity and is characterized by a defect rate per area.

10.7 Problems

1. You are working at a fabless company, and need to find a foundry to make 10,000 chips. You talk to two potential vendors.

   Company A says that they are using a slightly older, more mature, technology on 200 mm diameter wafers, with a defect density of only 0.2 cm^{-2}. You estimate you will need 0.5 cm^2 of area to make a single die using their process. The initial photomask cost (one-time cost) is $100,000. The cost per wafer is $2,000.

Company B is using a newer process (2 generations newer: length scales 0.7x per generation) with 300 mm diameter wafers. However, their defect rate is 10x higher than for company #2, the wafer cost is 3x higher, and the initial mask cost is 3x higher.

A) Fill in the following table.

	Company A	Company B
Photomask cost		
Wafer diameter		
Wafer cost		
Defect density		
Die Area		
# die / wafer		
Yield		
# good chips / wafer		
# wafers required		
Total cost (wafer + photomask)		
Cost per chip		

B) Which vendor would you choose for the initial run?

After the initial run, your company will determine if the sales meet expectations. If the product is successful, you will need to produce 250,000 chips per year.

C) Make a new table using 250,000 chips per year. Keep in mind, if you keep the same vendor, you do not need to pay for the photomask cost because they can be re-used. Thus, enter '0' for the photomask cost for whichever vendor you chose in part B. What is the cost per chip for each company? Which vendor would you choose to produce the 250,000 chips per year?

11 Small MOSFETs

When MOSFETs are made very small, they no longer behave as simply as they were described in Chapter 5. They are typically called short-channel MOSFETs because the small distance between the source and drain can cause many problems in their design. When the gate length is less than 1 μm, the MOSFET is called a sub-micron MOSFET. When the gate length is below 100 nm, they are called sub-100 nm MOSFETs.

11.1 Substrate Bias or Body Effect

This book covers the MOSFET (Chapter 5) before the p-n junction (Chapter 7). Therefore, the effects of the p-n junction on the MOSFET could not be discussed in Chapter 5. Let's revisit the MOSFET. There are two p-n junctions: from the source to the substrate, and from the drain to the substrate (Fig. 11.1). These two p-n junctions **MUST** be reverse biased or else current will flow regardless of the gate voltage. The source and drain are identical from a manufacturing point of view. From a circuit point of view, the convention is that the drain is biased more positive than the source in an NMOS, and the drain is more negative than the source in a PMOS. Finally, to keep all the p-n junctions reverse biased, the body must be biased negative with respect to the source in an NMOS, and the body must be biased positive with respect to the source in a PMOS.

	NMOS	PMOS
Drain Voltage	$V_D > V_S$ (drain is more positive)	$V_D < V_S$ (drain is more negative)
Body Voltage (to guarantee reverse bias)	$V_B \leq V_S$ (body is more negative) $V_{BS} \leq 0$ V	$V_B \geq V_S$ (body is more positive) $V_{BS} \geq 0$ V

In Chapter 5, we always assumed that both the source and the substrate were grounded. This is a good assumption when using discrete MOSFETs because the source and body electrodes are typically tied

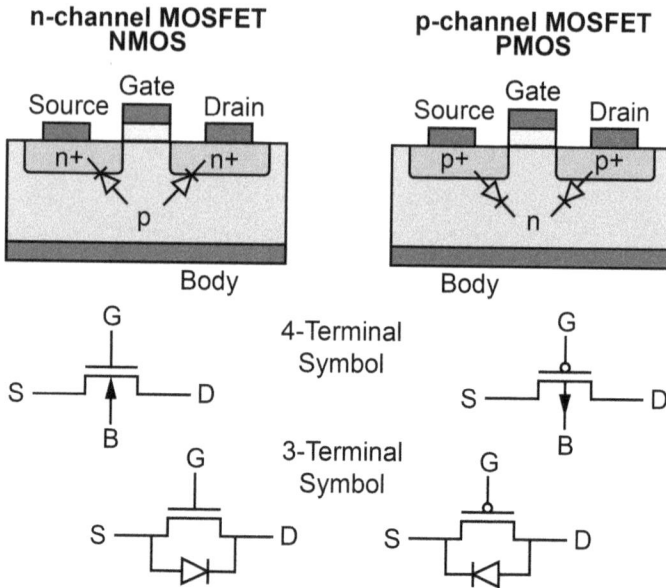

Figure 11.1: NMOS and PMOS transistors with the internal p-n junctions indicated as diodes. The 4-terminal symbol is appropriate when the body connection is available. The 3-terminal symbol is appropriate when the source and body are internally tied together.

together in the package and there are three pins on the MOSFET. Because the source and body are shorted together, the source-substrate p-n junction is shorted and can be eliminated from a schematic. This leaves just a single diode due to the drain-substrate p-n junction. Thus, a datasheet for a discrete MOSFET will typically show a diode in parallel with the MOSFET going from the source to drain (Fig. 11.1, 3-terminal symbol) called a **body diode**.

When designing ICs, however, it is very common for two MOSFETs in series to have a common substrate connection. One source is connected to the substrate (ground) and one source is connected to the drain of the other MOSFET. This may be seen in the example circuit shown in Figure 5.4.

We need to understand what happens when the source and body are not at the same voltage. The equations derived in this book are valid if we keep the source as the reference voltage and make a few modifications to the threshold voltage V_T when $V_{BS} \neq 0$ V. The equation for the depletion charge Q_D, from Equations 5.22, and 5.23, should be changed as follows:

$$Q_D = -\sqrt{2q\epsilon_s\epsilon_0 N_A(\phi_{SB} - V_{BS})} \qquad \text{(NMOS)} \qquad (11.1)$$

$$Q_D = \sqrt{2q\epsilon_s\epsilon_0 N_D(|\phi_{SB}| + V_{BS})} \qquad \text{(PMOS)} \qquad (11.2)$$

Be careful of the minus signs. The magnitude of the depletion charge always increases when a body voltage is applied. This value for Q_D is then used in Equation 5.27 to calculate a new threshold voltage. Since many MOSFETs are connected with the source and substrate shorted, it is convenient to keep the definition of the threshold voltage the same and find the change in the threshold voltage as a function of the substrate bias. The change in the threshold voltage is:

$$\Delta V_T = \frac{\sqrt{2q\epsilon_s\epsilon_0 N_A}}{C_i}\left[\sqrt{\phi_{SB} - V_{BS}} - \sqrt{\phi_{SB}}\right] \qquad (11.3)$$

$$\text{(NMOS)}$$

$$\Delta V_T = -\frac{\sqrt{2q\epsilon_s\epsilon_0 N_D}}{C_i}\left[\sqrt{|\phi_{SB}| + V_{BS}} - \sqrt{|\phi_{SB}|}\right] \qquad (11.4)$$

$$\text{(PMOS)}$$

For the NMOS, the threshold voltage becomes more positive. For a PMOS, the threshold voltage becomes more negative. Thus, **the effect of a substrate bias is to make it harder to turn a MOSFET on**, and thus a substrate bias reduces the drain current.

Later in this chapter, we will see that a larger V_T reduces the amount of leakage current in the MOSFET. Therefore, a circuit-design trick that can be used is to set $V_{BS} = 0$ V when the circuit must run fast, but to bias the substrate to obtain $V_{BS} > 0$ V for the NMOS when the circuit can run slower. This will save a lot of power when the circuit is in the "slow" mode. This can be useful for battery-powered applications, such as a cell phone or laptop.

Example 11.1: In Example 5.3 we found that the threshold voltage is 0.386 V. Let us find the new threshold voltage if $V_B = -5$ V.

Answer: Using Equation 11.3:

$$\Delta V_T = \frac{\sqrt{2q\epsilon_s\epsilon_0 N_A}}{C_i}\left[\sqrt{\phi_{SB} - V_{BS}} - \sqrt{\phi_{SB}}\right]$$

$$= \frac{\sqrt{2(1.6 \times 10^{-19}\ C)(11.8)\left(8.854 \times 10^{-14}\frac{F}{cm}\right)(10^{17}\ cm^{-3})}}{345\frac{nF}{cm^2}} \times$$

$$\left[\sqrt{0.834\ V - (-5\ V)} - \sqrt{0.834\ V}\right]$$

$$\Delta V_T = 0.796\ V$$

We see that the threshold voltage increases from 0.386 V to 1.182 V.

Example 11.2: In Example 5.5 we found that the threshold voltage to be -1.221 V for a PMOS. Let us find the new threshold voltage if $V_B = 5$ V.

Answer:

$$\Delta V_T = -\frac{\sqrt{2q\epsilon_s\epsilon_0 N_D}}{C_i}\left[\sqrt{\phi_{SB} + V_{BS}} - \sqrt{\phi_{SB}}\right]$$

$$= -\frac{\sqrt{2(1.6 \times 10^{-19}\ C)(11.8)\left(8.854 \times 10^{-14}\frac{F}{cm}\right)(10^{17}\ cm^{-3})}}{691\frac{nF}{cm^2}} \times$$

$$\left[\sqrt{0.834\ V + 5\ V} - \sqrt{0.834\ V}\right]$$

$$\Delta V_T = -0.398\ V$$

We see that the threshold voltage changes from -1.1221 V to -1.50 V.

11.2 Channel Length Modulation

Figure 11.2 shows a cross-section of a MOSFET with the depletion regions around the source and drain indicated. The channel length, L, was a fixed distance in Chapter 5 and equal to the gate length. We must revisit the concept of the channel length in this section. When designing a VLSI circuit using a CAD program, the **physical gate length** is drawn, and this distance becomes the physical separation of the source and drain doped regions. Hence, the physical distance between the source and drain is commonly called L_{drawn}, as indicated in Fig. 11.2. However, the electrons do not need to travel this entire distance. They only need to travel from one depletion region to the other depletion region. Therefore, the **effective channel length**, L, is the distance between the depletion regions. From the figure, we can calculate L by:

$$L = L_{drawn} - x_{P,S} - x_{P,D} \qquad \text{(NMOS)} \qquad (11.5)$$

where x_P is the extent of the depletion region into the p-type substrate due to the source and drain p-n junctions. Equation 7.12 with $V_{pn} = 0$ may be used to find $x_{P,S}$ since the source and substrate are both grounded. Note that the drain usually has a voltage applied to it.

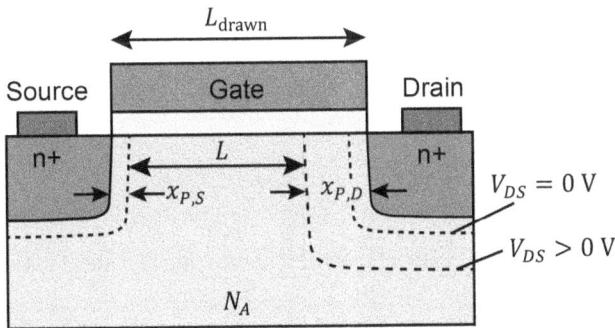

Figure 11.2: Schematic of an NMOS transistor showing the reduction in the effective channel length due to the depletion region from the source-substrate p-n junction and the drain-substrate p-n junction. When a drain voltage is applied, the drain p-n junction becomes larger, further reducing the channel length. The depletion region under the gate is not shown in this figure.

Therefore, to find $x_{P,D}$, we use $V_{pn} = V_B - V_D = 0 - V_{DS} = -V_{DS}$ in Equation 7.12. That is,

$$x_{P,S} = \sqrt{\frac{2\epsilon_s\epsilon_0}{qN_A}\left(\frac{N_D}{N_D + N_A}\right)(V_{bi})} \tag{11.6}$$

$$x_{P,D} = \sqrt{\frac{2\epsilon_s\epsilon_0}{qN_A}\left(\frac{N_D}{N_D + N_A}\right)(V_{bi} + V_{DS})} \tag{11.7}$$

There are a couple of important things to note. First, the depletion region width on the source side does not change because the source voltage is always zero. Second, the depletion region width on the drain side depends upon the drain voltage. Third, the effective channel length, L, depends on V_{DS} and is not a constant.

Similarly, for a PMOS transistor, we find:

$$L = L_{drawn} - x_{N,S} - x_{N,D} \qquad \text{(PMOS)} \tag{11.8}$$

$$x_{N,S} = \sqrt{\frac{2\epsilon_s\epsilon_0}{qN_D}\left(\frac{N_A}{N_D + N_A}\right)(V_{bi})} \tag{11.9}$$

$$x_{N,D} = \sqrt{\frac{2\epsilon_s\epsilon_0}{qN_D}\left(\frac{N_A}{N_D + N_A}\right)(V_{bi} - V_{DS})} \tag{11.10}$$

The fact that the effective channel length varies with the drain voltage is called **channel length modulation**. In an NMOS, an increase in the drain voltage results in a decrease in the effective channel length, and an increase in the drain current. In a PMOS, a more negative drain voltage results in a more negative (larger magnitude) drain current. An increase in the magnitude of the drain voltage results in a decrease in the effective channel length, and an increase in the magnitude of the drain current.

Figure 11.3 illustrates the effect of channel length modulation. The dotted lines show the drain current for no channel length modulation ($L = L_{drawn}$). The solid lines show the drain current with channel length

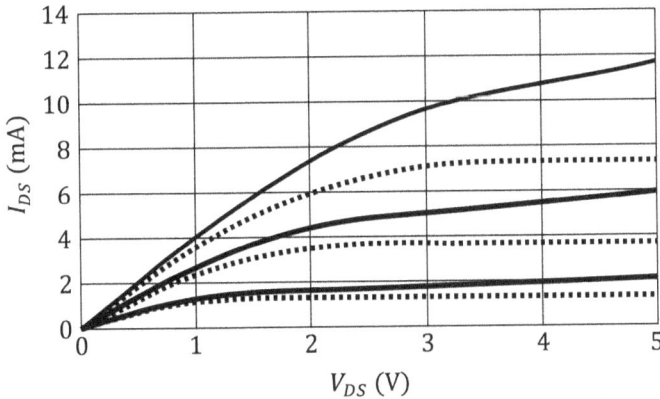

Figure 11.3: I-V curve for an NMOS. The solid curves show the drain current with channel length modulation, while the dotted curves show the drain current without channel length modulation.

modulation. The drain current no longer saturates, and the overall current is larger. For digital circuits, a little bit of channel length modulation may be a good thing because a larger drain current makes the digital circuits faster. For analog circuits, the channel length modulation may be a problem because it increases the output conductance.

Looking at Figure 11.3, we see that the drain current appears to increase linearly in the saturation regime. In reality, the slope isn't perfectly linear, but a linear approximation is often pretty good, especially for MOSFETs with a long channel for which the percent change in channel length is small over the voltage range of interest.

Example 11.3: Consider an NMOS with a substrate doping concentration of 2×10^{16} cm^{-3}. The source and drain are doped at a concentration of 10^{19} cm^{-3}. The gate length is 1 μm. What are (a) the effective channel length with no voltage applied, and (b) the effective channel length for $V_{DS} = 5$ V?

Answer: (a) The built-in potential for the p-n junctions is:

$$\phi_{bi} = 0.0259 \ln \frac{2 \times 10^{16} \text{ cm}^{-3} \cdot 10^{19} \text{ cm}^{-3}}{10^{20} \text{ cm}^{-6}} = 0.913 \text{ V}$$

Using Equation 11.6 and using the approximation $N_D \gg N_A$:

$$x_{P,S} = \sqrt{\frac{2 \cdot 11.8 \cdot 8.854 \times 10^{-14} \text{F/cm} \cdot 0.913 \text{ V}}{1.6 \times 10^{-19} \text{C} \cdot 2 \times 10^{16} \text{ cm}^{-3}}} = 244 \text{ nm}$$

We also find that $x_{P,D} = 244$ nm since the drain voltage is zero. Using Equation 11.5, the effective gate length is:

$$L = 1000 \text{ nm} - 244 \text{ nm} - 244 \text{ nm} = 512 \text{ nm}$$

(b) With $V_{DS} = 5$ V:

$$x_{P,D} = \sqrt{\frac{2 \cdot 11.8 \cdot 8.854 \times 10^{-14} \text{F/cm} \cdot (0.913\text{V} + 5\text{V})}{1.6 \times 10^{-19} \text{C} \cdot 2 \times 10^{16} \text{ cm}^{-3}}} = 621 \text{ nm}$$

The effective gate length is:

$$L = 1000 \text{ nm} - 244 \text{ nm} - 621 \text{ nm} = 135 \text{ nm}$$

With a bit more drain voltage, there wouldn't be a channel region at all!

To avoid having to calculate the effective channel length for every drain voltage, a commonly used model is to multiply the equation for the drain current from Chapter 5 by $1 + \lambda V_{DS}$, where λ is found by fitting a best fit to the actual MOSFET current in saturation. All the drain current equations in Table 5.1 are multiplied by $(1 + \lambda V_{DS})$. For example, the drain current equation in saturation for an NMOS becomes:

$$I_{DS} = \frac{1}{2} \mu_n C_i \frac{W}{L} (V_{GS} - V_T)^2 (1 + \lambda V_{DS}) \tag{11.11}$$

11.3 Output Conductance

In the last section, we saw that the drain current is not constant in saturation. Output conductance is defined to be:

$$g_0 = \frac{\partial I_{DS}}{\partial V_{DS}} \tag{11.12}$$

The output conductance in saturation was found to be zero in Chapter 5, but that only applies in an ideal case. The output conductance is clearly not zero for small MOSFETs, as can be seen from the fact that the current has a slope in saturation in Figure 11.3. If we insert Equation 11.11 into Equation 11.12, we get:

$$g_0 = \frac{\partial I_{DS}}{\partial V_{DS}} = \lambda I_{DS}\Big|_{sat} \tag{11.13}$$

The non-zero output conductance is a big deal for many analog circuits. Transistors for analog circuits often use longer gate lengths than transistors in digital circuits to obtain a lower output conductance.

11.4 Sub-threshold Current

Ideally, when a MOSFET is turned off, zero current flows. However, the MOSFET is not completely off. The source to substrate, and the drain to substrate, are reverse biased diodes, and we already know that a small amount of current flows in reverse bias. If we look at a diagram of the MOSFET showing the depletion regions, such as shown in Figure 11.4, we see that the two heavily doped n- regions making up the source and drain are connected by a depletion region under the gate. Figure 11.4 is shown for strong inversion, but the depletion regions connect the source and drain even in the depletion and weak inversion modes. Only in accumulation is there no depletion region under the gate. Remember also that a depletion region doesn't mean there are no electrons or holes; a depletion region means that the quantity of electrons and holes are sufficiently small that the fixed dopant atoms determine the charge in this region. Thus, a small amount of current can flow, but since the

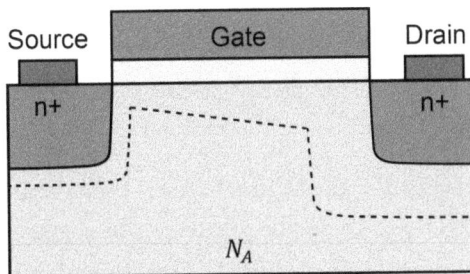

Figure 11.4: Illustration showing that there is a depletion region connecting the source and drain.

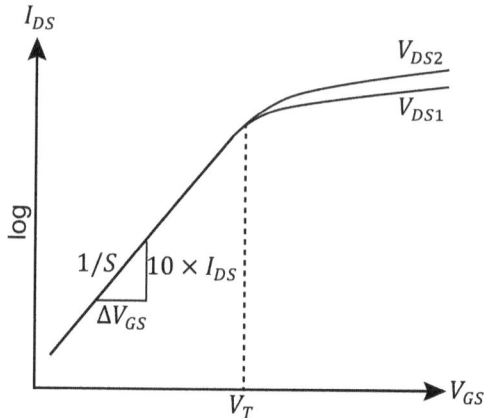

Figure 11.5: NMOS drain current versus gate voltage on a log scale, showing the sub-threshold current that flows when $V_{GS} < V_T$.

concentration of electrons in the depletion region varies exponentially with the gate voltage, the current flow varies exponentially with the gate voltage.

Figure 11.5 shows the drain current versus gate voltage on a log scale for an NMOS transistor. There can be a significant amount of current when the transistor is off. The current that flows when the gate voltage is below the threshold voltage is called the **sub-threshold current**. The sub-threshold current decreases very rapidly as the gate voltage is reduced below the threshold voltage. Above the threshold voltage the drain current appears to increase gradually with gate voltage, as a log scale makes a parabolic shape look similar to a gradual slope. The drain current is shown for two different drain voltages, resulting in two different drain currents when the MOSFET is on. However, the drain current is the same when the MOSFET is off, regardless of the drain voltage.

Since the sub-threshold current appears to be a straight line on a log scale, it is convenient to describe the sub-threshold current using this slope, or rather the inverse of this slope as it appears in Figure 11.5. The **sub-threshold slope**, S, is defined as the change in gate voltage required to obtain a 10x change in drain current:

$$S = \frac{\Delta V_{GS}}{10 \times \Delta I_{DS}} = \ln(10)\frac{kT}{q}\left(1 + \frac{C_D}{C_i}\right) \qquad (11.14)$$

where the first definition is useful for reading the slope from measured data, and the second definition is useful for calculations. $C_D = \epsilon_s\epsilon_0/W_{dm}$. W_{dm} and C_i can be found from Equations 5.20 and 5.26.

We want as small a value of S as possible to get the steepest slope for the drain current. This in turn minimizes the sub-threshold current when the gate voltage is below the threshold voltage. The theoretical best value for S, for a conventional MOSFET, is 60 mV / decade I. In practice values of 70 or 80 mV / decade I are often obtained, and MOSFETs are often considered "poor" if the sub-threshold slope exceeds 100 mV /decade I.

Example 11.4: Consider a MOSFET with the following I-V curve. What are (1) the threshold voltage and (2) the sub-threshold slope?

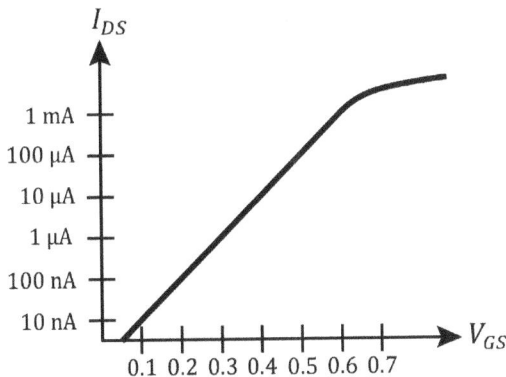

Answer: (a) The threshold voltage is approximately 0.6 V because that is where the slope of the sub-threshold current drops. (b) Pick two currents that are 10x apart, say 10 nA and 100 nA. The difference in voltage is 100 mV. Therefore, $S = 100$ mV / decade I.

Example 11.5: Consider a 500 million transistor digital circuit with a 1.5 V power supply. The threshold voltage is 0.4 V. Ignoring any noise in the circuit, we assume that the low voltage is 0 V. On average, half of the transistors in the circuit are turned on and half of them are turned off. The current that flows at the threshold voltage is 10 μA, and the sub-threshold slope is 100 mV / decade I. How much power is dissipated by the transistors that are "off".

Answer: Going from 0.4 V to 0 V, there are 4 sets of 100 mV. Therefore, the current is reduced by a factor of 10^4. That is, the sub-threshold current is $I = 10$ μA $\times 10^{-4} = 1$ nA. Multiplying by 250 million transistors, the total current is $I = 250$ mA. Multiplying by the supply voltage, the total power dissipated by the "off" transistors is 375 mW.

This example shows that power can be wasted by the transistors that are supposed to be off. 375 mW is not much for many applications, but it is substantial for battery operated devices.

Example 11.6: How long would the battery on an Apple iPhone 5 last for the circuit in Example 11.5, assuming that the battery was devoted to supplying this chip with its required "wasted" power, and no other circuits were enabled on the phone. The battery size is 5.45 Wh.

Answer:

$$t = \frac{5.45\ Wh}{375\ mW} = 14.5\ hr$$

That is, a high-density lithium ion battery can only power the "wasted" power from this one IC for more little more than half a day.

Example 11.7: Consider the circuit from Example 11.5, but to reduce power consumption the designers decreased the supply voltage to 1 V and increased the threshold voltage to 0.6 V. How long will the lithium ion battery from Example 11.6 last in this case?

Answer:

$$P = IV = (10 \ \mu A \times 10^{-6})(250 \times 10^6)(1 \ V) = 2.5 \ mW$$

$$t = \frac{5.45 \ Whr}{2.5 \ mW} = 2180 \ hr$$

This far exceeds the battery life of the iPhone during normal usage, and thus this "wasted" power may be ignored. Notice that the change that caused the greatest reduction in power is to increase the threshold voltage, which has the side-effect of slowing down the digital circuit due to the lower current.

The examples above showed that a small leakage current can be significant for portable applications. In some applications this leakage current is negligible. For example, a CPU used for a workstation may dissipate 100 W of power. An additional 250 mW is a trivial amount of power compared to the total power dissipated.

One commonly used equation for the sub-threshold current is:

$$I_{DS} = \mu_n C_i \frac{W}{L} \left(\frac{k_B T}{q}\right)^2 \exp\left(\frac{q(V_{GS} - V_T)}{m k_B T}\right)\left[1 - \exp\left(-\frac{q V_{DS}}{k_B T}\right)\right]$$
$$(V_{GS} < V_T), \quad \text{NMOS} \tag{11.15}$$

where

$$m = 1 + \frac{1}{C_i}\sqrt{\frac{\epsilon_s \epsilon_0 q N_A}{2\phi_{SB}}} \tag{11.16}$$

There are some interesting things about this equation. First, the drain current increases exponentially with gate voltage. Second, the drain current is independent of drain voltage for drain voltages larger than 0.1 V.

11.5 Gate Current

Using quantum mechanics, it can be shown that a particle can pass through a barrier. Electrons are capable of moving through an insulator. First the electron is on one side of the insulator, then it is on the other side of the insulator. This is called **tunneling** through the barrier.

An analogy would be if I were walking down the street, going to my office. To get in, I avoid the doors and simply walk next to the wall of my office. First, I am on the outside of a building. Then, suddenly, I am on the inside of the building, on the other side of the wall. That would be a pretty amazing trick, and quantum mechanics says it can happen. However, the odds of that happening are so small that even if everyone on this planet had stood next to a wall when the big bang occurred, we would all still be outside today and not one of us would have "tunneled" through the wall to the inside.

An electron, with a tiny mass and small size, has a reasonable chance of tunneling through a very thin insulator. If the insulator thickness is less than 5 nm, there is a possibility for the electron to tunnel through the barrier. If the distance is greater than 5 nm, the chances are usually negligible.

Ideally, no static current flows through the gate of a MOSFET. A dynamic current, $i = C_G \, dV_G/dt$ does flow as the gate voltage changes because of the gate capacitance. But no current should flow for a constant gate voltage. This is one of the reasons MOSFETs are used in today's integrated circuits. But the need to make very thin gate oxides to get a good drain current and minimize DIBL (described later) has resulted in a substantial amount of gate current due to tunneling. There are two types of tunneling mechanisms to consider. (1) Electrons may tunnel from the gate electrode, through the gate insulator, and into the channel region. This is called **direct tunneling**. (2) Electrons may tunnel from the gate electrode into the conduction band of the gate insulator. This process is known as **Fowler-Nordheim tunneling**. Let us consider each type of tunneling.

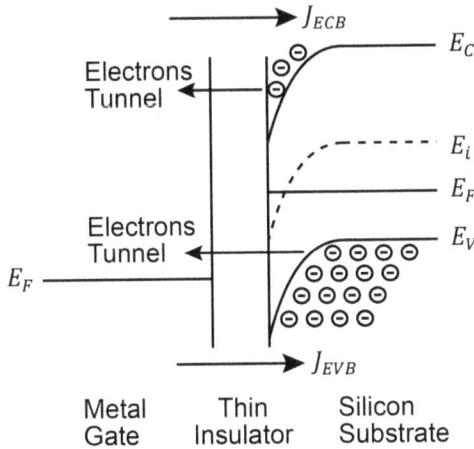

Figure 11.6: Energy band diagram of an NMOS showing the gate current density due to electrons tunneling from the conduction band, J_{ECB}, and due to electrons tunneling from the valence band, J_{EVB}.

11.5.1 Direct tunneling through the gate insulator

A band diagram of the MOSFET with a thin insulator in strong inversion is shown in Figure 11.6. Electrons may tunnel from the conduction band, which has a lot of electrons because the MOSFET is in strong inversion, and electrons may tunnel from the valence band, which is nearly full of electrons. It depends on the exact alignment of the energy band, but the electrons in the conduction band usually have a place to tunnel to in the gate, whether the gate is a metal or polysilicon. Electrons in the valence band, which has a lot more electrons than the conduction band, may or may not be able to tunnel into the gate, depending upon the voltage and the density of empty quantum states that the electron may tunnel into. Keep in mind that tunneling does not involve a change in energy, so the electron can only move horizontally in Figure 11.6.

Electrons tunneling from the substrate to the gate creates current flow from the gate to the substrate. The component of the current density that flows from the conduction band is labeled J_{ECB} for electron current density to/from the conduction band. The component of the current

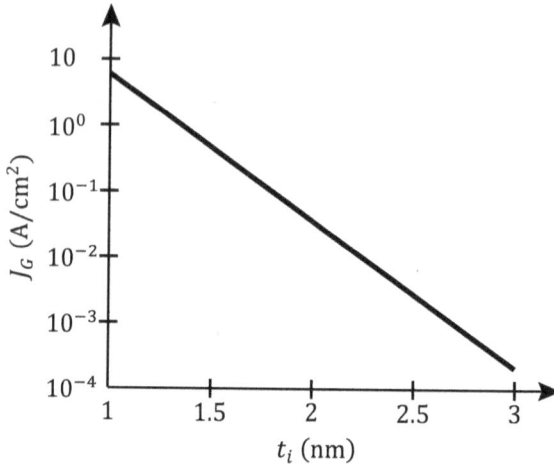

Figure 11.7: Typical gate tunneling current density as a function of the insulator thickness for $V_{GS} = 1\,V$. The gate material and the substrate doping concentration have a large effect on the tunneling current, so this graph can only be used as a guide.

density that flows from the valence band is labeled J_{EVB} for electron current density to/from the valence band.

The gate insulator has been made very thin in recent years; as thin as 1.2 nm. When the gate insulator is very thin, the tunneling rate may be very high. Figure 11.7 shows a typical gate current density as a function of the insulator thickness. The tunneling current varies exponentially with the insulator thickness, creating a strong desire to keep the insulator thickness as thick as possible.

The gate current may flow directly into the source and drain at either end of the channel, but most of the gate current flows directly into the channel region. If there is an inversion layer, the gate current will combine with the drain current and be captured as source current. If there is no inversion layer, the gate current will go directly into the substrate and becomes substrate current.

Example 11.8: Consider an IC that measures 0.5 cm x 0.5 cm and uses a 1 V power supply. Assume approximately 10% of the area of the IC has a gate on it, and only ½ of the transistors have a voltage on the gate. The insulator thickness is 2 nm. How much gate current flows? Repeat for an insulator thickness of 1.5 nm.

Answer:

From Fig. 11.7, the current density is approximately 0.05 A / cm^2 for a 2 nm insulator thickness. The gate current is then:

$$I_G = 0.05 \; \frac{A}{cm^2} \times 0.5 \text{ cm} \times 0.5 \text{ cm} \times 0.1 \times 0.5 \approx 600 \; \mu A$$

If the insulator is reduced to 1.5 nm, the current density increases a bit more than 10x. Thus, $I_G \approx 6$ mA.

11.5.2 Fowler-Nordheim Tunneling

Electrons don't have to tunnel directly into the gate from the substrate. They may also tunnel into the conduction band of the insulator, and from there move into the gate. After all, an insulator is nothing more than a semiconductor with a very large band gap that results in a nearly zero electron concentration in the conduction band. If electrons tunnel into the conduction band of the insulator, they are free to move. The resulting current, called **Fowler-Nordheim current**, is labeled J_{FN}. An energy band diagram illustrating Fowler-Nordheim tunneling is in Figure 11.8.

Fowler-Nordheim tunneling may occur for insulator thicknesses greater than 5 nm because the effective thickness of the insulator thickness for tunneling to occur becomes smaller once the triangular region of the band diagram extends down to the conduction band of the semiconductor. As the energy bands bend more, the effective thickness that the electron have to tunnel through becomes smaller. For a large enough voltage, Fowler-Nordheim tunneling can even occur for thick insulators.

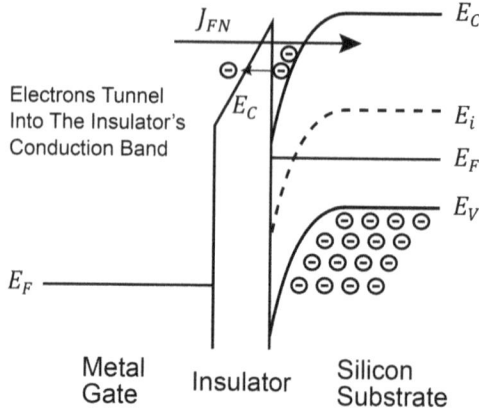

Figure 11.8: Energy band diagram showing electrons tunneling from the conduction band of the semiconductor into the conduction band of the insulator. This is called Fowler-Nordheim tunneling.

A commonly used equation for Fowler-Nordheim tunneling is:

$$J_{FN} = \frac{q^3 \mathcal{E}_{ox}^2}{16\pi^2 \hbar \phi_{ox}'} \exp\left(-\frac{4\sqrt{2m^*}\,\phi_{ox}'^{3/2}}{3\hbar q \mathcal{E}_{ox}}\right) \tag{11.17}$$

where $\mathcal{E}_{ox}$ is the electric field in the oxide, $\hbar$ is the reduced Plank's constant, and ϕ_{ox}' is the energy barrier between the conduction band of the oxide and the silicon.

11.6 Strained Silicon

When silicon is mechanically strained, the energy bands change. The band gap changes and, most importantly, the mobility of the electrons and holes change. This can be used to advantage to significantly improve the performance of MOSFETs. In fact, the electron mobility and hole mobility may be more than doubled using strain.

If an NMOS transistor has a uniaxial tensile strain in the channel, the mobility of the NMOS transistor is improved. A tensile strain means that the atoms are stretched further apart than normal, and uniaxial strain means that the atoms are only stretched apart in the direction of the channel from the source to the drain.

If a PMOS transistor has uniaxial compressive strain in the channel, the mobility of the PMOS transistor is improved. A compressive strain means that the atoms are pushed closer together than normal. The uniaxial strain must be in the direction of the channel from the source to the drain.

Strained silicon was first introduced to the manufacturing process at the 90 nm node and has been adopted by all the major manufacturing companies. It is easier to introduce the strain as the transistors become smaller, making this technique attractive for small transistors. The performance improvement for a PMOS transistor is larger than that for an NMOS transistor.

The type of strain (compressive or tensile), direction of strain, and the orientation of the channel on the surface, and the crystal face used to make the MOSFET all affect the final mobility and affect the NMOS and PMOS differently.

11.7 Metal Gate / High-k Dielectric

To make transistors faster, the gate capacitance, C_i is made larger. The equation for C_i is:

$$C_i = \frac{\epsilon_i \epsilon_0}{t_i} \tag{11.18}$$

Traditionally the gate insulator is SiO_2 and C_i is improved by shrinking the insulator thickness. However, the tunneling current becomes extremely large through a thin gate insulator. To solve this problem, it is desired to use an insulator thickness sufficiently large to prevent tunneling, and increase the dielectric constant of the insulator to increase C_i. Some books use κ to represent the relative dielectric constant, and thus it is common to talk about high-k, for high-κ, for a gate insulator made with a material with a relative dielectric constant larger than that for SiO_2. HfO_2 is a commonly used high-k material, with $\epsilon_i = 25$.

The problem with high-k materials is that they look good on paper but experimentally, they have a lot of defects at the silicon-insulator interface and the carrier mobility is greatly reduced. To avoid the carrier mobility degradation, two things must be done. First, it is necessary to use a monolayer of SiO_2 over the silicon to minimize the number of interface defects. Second, a metal gate is required because the metal

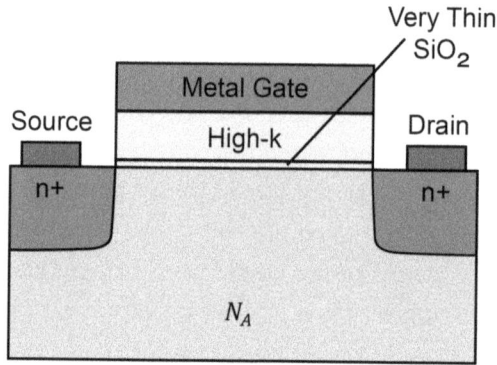

Figure 11.9: Cross-section of the metal gate / high-k gate dielectric MOS structure. At the time this is written, all high-k gates still contain a thin SiO₂ layer to minimize surface states at the top of the silicon.

gate screens elastic waves that scatter the carriers. Thus, when one talks about high-k gate dielectrics, it is also paired with metal gates. A schematic of the metal gate / high-k dielectric solution is shown in Figure 11.9.

Using a metal gate is an advantage from an IC performance point of view because it decreases the gate resistance, increasing speed and reducing power consumption. However, the NMOS and PMOS transistors require different metals for their gates, increasing the manufacturing costs.

Overall, the high-k / metal gate solution provides a significant increase in performance and a significant savings in power. The major IC manufacturers have adopted high-k / metal gates.

11.8 Hot Carrier Effects

The final section of this chapter will deal with reliability. In thermal equilibrium, the entire semiconductor is at a uniform temperature. Once a voltage is applied, the semiconductor is no longer in thermal equilibrium and different portions may be at different temperatures. Interestingly, the electrons may be at a different temperature than the atoms through which they pass. This is possible when a large electric field is present that accelerates the electrons, but not the atoms since the atoms are fixed in place. Energy may be related to the kinetic energy of the electrons and to temperature through the following equations:

$$E = \frac{1}{2}mv^2 \qquad (11.19)$$

$$E = k_B T \qquad (11.20)$$

From Equation 11.19, we can find an effective temperature for the electrons based on their average speed. From these equations, we can see that the electrons have a higher temperature when accelerated by an electric field. Since Equation 11.20 is a statistical equation, some electrons are "hotter" and some are "colder" than we would calculate using Equation 11.20. When we refer to a "hot" electron, we are really referring to an electron that has more energy than the average energy of the semiconductor. On an energy band diagram, a "hot" electron is no longer at the bottom of the conduction band; it is at a higher energy.

Hot electrons have the ability to jump into defects that are at an energy higher than the bottom of the conduction band. This is important for MOSFETs, where electrons that travel along the channel are easily "hot" due to the large electric fields present, and there is an insulating film adjacent to the channel. Some of the hot electrons get trapped in the insulator, causing several problems. First, the trapped electrons affect the threshold voltage of the MOSFET. Second, the trapped electrons are near the channel and deflect electrons traveling through the channel, reducing the effective mobility and thereby reducing the drain current. Third, these trapped electrons degrade the insulating properties of the insulator, resulting in breakdown of the insulator. Unlike p-n junctions, **the breakdown of an insulator is destructive**.

In summary, hot electrons are a problem for reliability. Before MOSFETs are ready for mass production, a manufacturer will study the reliability of the MOSFET and the failure mechanisms. If hot electrons are found to cause premature failure, then the MOSFET must be redesigned to limit the electric field in the channel.

In a MOSFET in strong inversion, the electric field is strongest next to the drain where the channel is pinched-off. To reduce the electric field, MOSFETs are sometimes made with a lightly doped drain region that

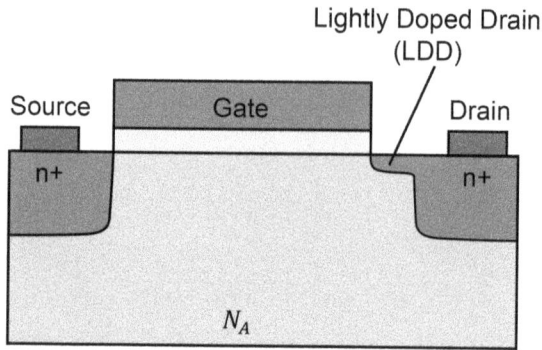

Figure 11.10: Schematic of a MOSFET that incorporates a lightly-doped drain (LDD).

increases the resistance near the drain and slows down the electrons. A schematic of a MOSFET with a **lightly-doped drain** (LDD) is shown in Figure 11.10. This solution to the hot electron problem has the tradeoff that it increases the channel resistance, lowers the drain current and slows down MOSFET circuits.

11.9 Summary of Key Concepts

- Traditionally transistors shrink by a factor of 0.7 every 2 years.
- Applying a voltage to the substrate increases the threshold voltage of an NMOS, making it harder to turn on. Similarly, the threshold voltage of a PMOS becomes more negative.
- Current flows through the drain even when the MOSFET is in cutoff. This is called the sub-threshold current.
- The sub-threshold current drops by 10x when the gate voltage drops by $60 - 100$ mV.
- Thin insulating gate layers permit a tunneling current to arise, which results in gate current.
- Direct tunneling is when electrons tunnel through a gate insulating layer.
- Fowler-Nordheim tunneling is when electrons tunnel into the conduction band of the gate insulating layer.
- MOSFET reliability is affected by electrons that gain a high velocity due to a high electric field. If the electron is at the semiconductor-insulator interface, these electrons may deflect

into the insulator, where they become trapped and affect the threshold voltage as well as the dielectric properties.

- All transistors that are manufactured with a gate length < 90 nm are now made using strain the channel region. The strain changes the mobility and improves MOSFET performance.
- Many MOSFETs are now made using a combination of a high-k dielectric and a metal gate. The high-k improves the gate capacitance, improving performance, and the metal gate reduces parasitic resistance.

11.10 Problems

1. Consider an NMOS structure with a p-type silicon substrate doped with $N_A = 10^{16}$ cm^{-3} boron atoms. The source and drain are doped n-type with $N_D = 10^{19}$ cm^{-3}. The drawn gate length is $L_{drawn} = 3\ \mu m$. The insulator is SiO$_2$ with a thickness of 50 nm. $N_{ss} = 2 \times 10^{10}$ cm^{-2}. The gate is a polysilicon gate that is heavily doped n-type. The electron mobility is $\mu_n = 350\ \frac{cm^2}{V \cdot s}$. $W = 10\ \mu m$. We will investigate the channel length modulation model in this problem.
   a. What is the flat band voltage?
   b. What is the threshold voltage?
   c. What is the effective gate length, L, with zero applied voltage?
   d. If the drain voltage is $V_{DS} = 5\ V$, what is the new effective gate length L ?
   e. Plot I_{DS} versus V_{DS} for $V_{DS} = 0 \cdots 5\ V$. Use $V_{GS} = 2V$. Assume that the gate length does not change.
   f. Now perform the same calculation as part (e), but vary the gate length with the drain voltage. Overlay this graph on the graph from part (e).

2. Consider the transistor from problem #1. A voltage of -5 V is applied to the substrate. What is the new threshold voltage?

3. For the MOSFET in problem #1, calculate the sub-threshold current for:
   a. $V_{GS} = 0.25\ V$ and $V_{DS} = 1\ V$.
   b. $V_{GS} = 0\ V$ and $V_{DS} = 3\ V$.

4. For the MOSFET in problem #1, plot I_{DS} versus V_{GS} for $V_{GS} = 0 \cdots 3\,V$. Use $V_{DS} = 3\,V$. Include the sub-threshold current.
   a. Use a linear scale for current.
   b. Use a log scale for current.

   Note: It is OK if the curves for the sub-threshold current and the "normal" current don't meet perfectly. Can you explain why?

5. Repeat problem #4 with gate length modulation. On a single plot, show the results for (a) no gate length modulation, and (b) gate length modulation.

6. Using your plot from problem #4, what is the sub-threshold slope S?

7. Using Equation 11.14, what is the sub-threshold slope S?

8. Repeat problem #1, but start with a gate length of 1.2 µm. Limit the maximum voltage on the plots to 3 V.

9. Draw an energy band diagram similar to Figure 11.7, but for an n-type substrate. A negative voltage is applied to the gate. Indicate how the electrons tunnel for (a) direct tunneling and (b) Fowler-Nordheim tunneling.

10. Consider the transistor from problem #1. The SiO_2 gate insulator is replaced with HfO_2, which has a relative dielectric constant of 25. By what factor does the drain current increase?

12   Very Small MOSFETs

I wanted to call this chapter, "Why won't my transistor turn off!" but that didn't seem like the proper name of a book chapter. In this chapter we will see that everything seems to conspire to allow current to flow from the drain to the source. The drain voltage reduces the threshold voltage, current flows through the reverse biased p-n junctions, and quantum tunneling occurs!

In this chapter we will see why it is difficult to turn off very small MOSFETs, and how we solve the problem. In particular, we will end with a discussion of the FinFET and the Gate All Around (GAA) MOSFET.

12.1   When is a MOSFET OFF?

If we apply $V_{DS} > 0$ V, there is always current flow from the drain to the source no matter the gate voltage. So what do we mean for a MOSFET to be off?

An easy definition is that the MOSFET is off when $V_{GS} < V_T$. This was the definition used in Chapter 5. It is a commonly used definition when we do not need to worry about a small amount of sub-threshold current.

But there is always some amount of current that flows, called the leakage current. This current can be rather large. So perhaps a better definition would be: the MOSFET is off when $V_{GS} < V_T$ and the leakage current is tolerable.

12.2   Drain Induced Barrier Lowering (DIBL)

In a very small MOSFET, the source-substrate and the drain-substrate p-n junctions have a strong influence on the electrostatics in the channel region of the MOSFET. Figure 12.1 shows a cross-section of a MOSFET with the depletion regions indicated. There are dotted lines to try to delineate the portion of the depletion region that belongs to the source, the portion that belongs to the drain, and the portion that belongs to the gate. It is called a trapezoidal approximation when the depletion charges are separated in this manner. It is not mathematically rigorous, but it does a good job at explaining the effects of the shared depletion regions and is easy to understand. Consider the equation for the threshold voltage. The last term in Equation 5.27 is:

$$V_T = \cdots - \frac{Q_D}{C_i}$$

$$(12.1)$$

When the surface depletion charge due to the gate, Q_D, is shared with the source and drain, the magnitude of Q_D decreases. In an NMOS, Q_D is negative and sharing the charge will make Q_D less negative and therefore the threshold voltage will become more negative.

Wow... all those negative signs are confusing. Let us consider an example with made-up, unrealistic numbers. Let $Q_D = -5$ because the depletion charge is negative. Then Equation 12.1 says $V_T = \cdots (-)(-5) = 5$. After sharing the charge, $Q_D = -4$ (smaller magnitude). Then Equation 11.14 says $V_T = \cdots (-)(-4) = 4$. Thus, the threshold voltage became smaller.

In a PMOS, Q_D is positive and sharing the charge will make Q_D less positive and the threshold voltage will become more positive. For both NMOS and PMOS, it becomes easier to turn on the transistor. In other words, **sharing the charge makes it harder to turn off the transistor.**

In Figure 12.1a, the gate length is large and the channel depletion charge is minimally affected by the source and drain. However, as the gate length becomes small (Figure 12.1b), the source and drain depletion regions "share" the depletion charge controlled by the gate. Let us define a modified channel depletion charge density, $\acute{Q}_D$. The total charge under the gate is the surface charge density multiplied by the area of the gate, WL. The total charge controlled by the gate (not the charge per area) is:

$$\acute{Q}_D WL = Q_D WL + \frac{qN_A}{2} WW_{dm}x_{P,S} + \frac{qN_A}{2} WW_{dm}x_{P,D}$$

$$(12.2)$$

(NMOS)

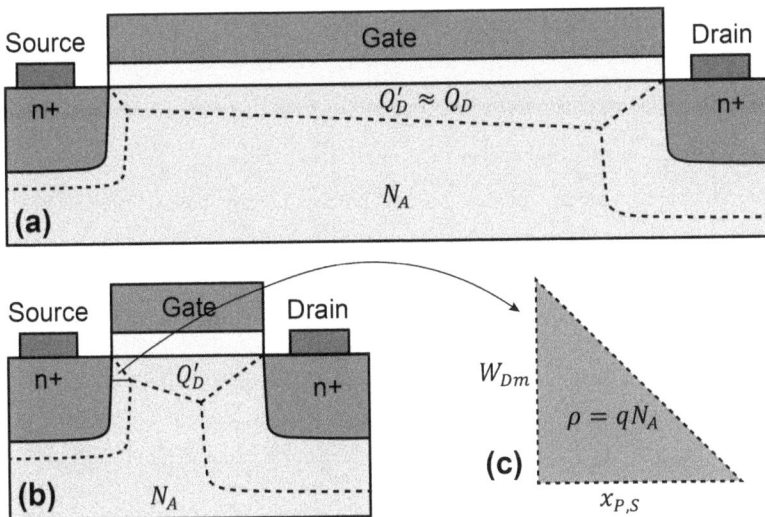

Figure 12.1: An NMOS in strong inversion with the depletion regions indicated. (a) Long gate length. Except end effects, the gate has complete control over the depletion charge along the channel. (b) Short gate length. The effective depletion charge controlled by the gate is greatly reduced due to charge sharing with the source and drain. (c) The source controls this triangular portion of the depletion charge.

where we have removed the triangular area s controlled by the source (Figure 12.1c) and drain. W_{dm} is the maximum depletion region width, as defined in Chapter 5, x_P is the depletion region width due to the p-n junction in the p- region (substrate). We are adding values to Q_D because in an NMOS Q_D is negative and we are reducing the magnitude. The depletion region width due to the source $(x_{P,S})$ will be different from that due to the drain $(x_{P,D})$ if a voltage is applied to the drain. Dividing by WL, we get:

$$\acute{Q}_D = Q_D + \frac{qN_A}{2L} W_{dm}(x_{P,S} + x_{P,D}) \qquad \text{(NMOS)} \qquad (12.3)$$

Similarly, for a PMOS:

$$\acute{Q}_D = Q_D - \frac{qN_D}{2L} W_{dm}(x_{N,S} + x_{N,D}) \qquad \text{(PMOS)} \qquad (12.4)$$

Looking at Figure 12.1a, we see that $|\acute{Q}_D| \cong |Q_D|$. In a MOSFET with a long channel, Q_D is hardly affected by the source and drain depletion regions. Now consider Figure 12.1b. In this diagram, the gate length is much shorter than Figure 12.1a. The modified gate depletion charge changes greatly $(|\acute{Q}_D| \ll |Q_D|)$ because x_P is a significant fraction of the gate length, L. This shows that the threshold voltage may be greatly reduced for MOSFETs with a short gate length compared to MOSFETs with a long gate length.

The problem is exacerbated when a voltage is applied to the drain. The drain voltage reverse biases the drain p-n junction, causing $x_{P,D}$ to become larger. If the gate length is small, then the change in the modified gate depletion charge can be significant. Thus, MOSFETs with small gate lengths have a threshold voltage that depends on the drain voltage. A large drain voltage will make it harder to turn off the MOSFET. Another effect of having a threshold voltage that depends on the drain voltage is that the drain current increases at larger drain voltages, which means the drain current is no longer constant in saturation. Not having the transistor turn off is a problem for digital circuit designers. Not having a constant saturation current is a problem for analog circuit designers who rely on this feature to implement good current sources, which are one of the basic building blocks to many analog circuits such as an integrated amplifier.

Since the drain voltage is changing the drain depletion region, this phenomenon is called **drain induced barrier lowering (DIBL)**, where the barrier that is lowered is the barrier to current flow when the MOSFET is off.

An accurate analysis is beyond the scope of this book. Although there are mathematical approaches available in the literature, the 2-D nature of this problem generally requires a numerical simulation to accurately predict the effects of DIBL. To avoid DIBL, it is common to increase the substrate doping concentration, which reduces x_P. Because this is inherently a 2-D problem, making the source and drain regions very thin also helps reduce DIBL. Unfortunately, these solutions involve a design tradeoff. A higher substrate doping concentration reduces the depletion layer width, increasing the gate capacitance and slowing down the

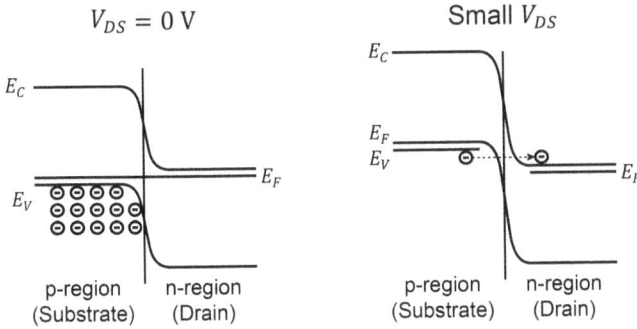

Figure 12.2: Illustration of band to band tunneling (BTBT). (a) With $V_{DS} = 0\,V$, the electrons in the valence band are almost at the same energy level as the conduction band in the drain. (b) Application of a small voltage on the drain permits the electrons to tunnel into the drain.

transistors. A thin source and drain increases the source and drain series resistance.

The DIBL effect becomes so severe for very small MOSFETs that the semiconductor industry switched to using a new type of MOSFET, called a FinFET, which will be described later in this chapter.

12.3 Band-to-band tunneling

As the source and drain of a MOSFET get close to each other, it is necessary to increase the doping concentration of the substrate to keep the depletion regions from interacting. This results in a very thin depletion region, and electrons may tunnel from the valence band of the substrate into the drain, and hence the name **band-to-band tunneling,** abbreviated as **BTBT**. This tunneling results in current flow from the drain to the substrate. This is the same physics as occurs in a Zener diode. Figure 12.2 illustrates this phenomenon. The valence band in the substrate is almost at the same energy as the conduction band in the drain when the doping concentration is very high (Figure 12.2a). With the application of a small drain voltage, which reverse biases the p-n junction, the valence band in the substrate is at a higher energy than the conduction band in the drain and electrons can tunnel across the barrier.

When we discussed Zener breakdown, we used the concept of a single breakdown voltage. In reality, the current has a non-linear relationship with respect to the reverse bias voltage across the p-n junction. As the reverse bias voltage increases, the number of electrons that can tunnel increases due to the increased energy overlap. Thus, the current increases rapidly. Tunneling is non-destructive.

Increasing the substrate doping concentration makes the depletion region smaller, reducing the drain voltage at which band-to-band tunneling starts. This makes it necessary to design the MOSFET with both the drain voltage and substrate doping concentration in mind to keep the band-to-band tunneling current to an appropriate level.

12.4 Silicon On Insulator (SOI) MOSFETs

Figure 12.3 shows the concept. The starting material is called a silicon on insulator (SOI) wafer. The MOSFETs are then formed in the top silicon layer. The silicon between the MOSFETs is etched away, leaving islands of silicon, each with a single MOSFET. These MOSFETs are more expensive to manufacture because a layer of SiO_2 is formed underneath a thin, perfect, single crystal of silicon.

The biggest advantage of an **SOI MOSFET** is that the leakage current from the drain to source is greatly reduced by physically removing the current path. Other advantages of using SOI MOSFETs for integrated circuits is that it makes it easy to electrically isolate all the MOSFETs from each other. Surrounding each MOSFET, the silicon is etched and SiO_2 is deposited. Now every transistor can have its own substrate and the substrate bias effect may be avoided, speeding up digital circuits. Other advantages include lower capacitance, the elimination of latch-up in CMOS circuits, and a better sub-threshold slope.

SOI MOSFETs are characterized as partially depleting or fully depleting, where the term is referring to the portion of the substrate that the depletion layer occupies. In a partially depleted SOI MOSFET, the depletion layer under the gate does not reach the bottom of the p-type substrate in strong inversion (NMOS). In a fully depleted SOI MOSFET, the depletion layer under the gate extends fully to the buried SiO_2 layer and there is no neutral p- region left. The smallest SOI MOSFETs are generally fully depleted.

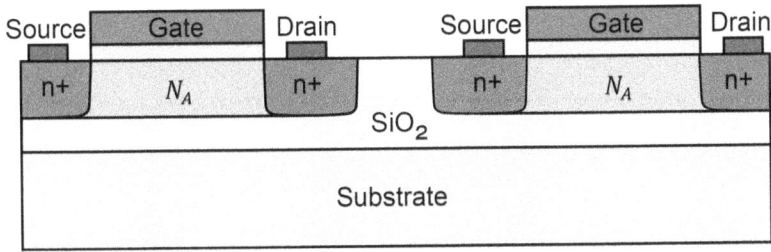

Figure 12.3: Schematic of two SOI MOSFETs that illustrates the electrical isolation between the MOSFETs.

The substrate acts like a 2nd gate on the bottom of the MOSFET. An SOI MOSFET can be thought of as a **dual gate MOSFET**. The channel region is shown in Figure 12.4. All of the bottom gates are connected since they are on a common substrate. Since we do not have individual control of the bottom gate, the FET has to be designed to ensure that the bottom gate does not turn the FET on. Just as there is a sub-threshold current due to the top gate, there is a sub-threshold current due to the bottom gate, although they are not independent of each other. Unfortunately, this is a 2D problem, and it requires solving Poisson's equation in 2D to determine the electric potential and charge distribution. This book will not try to derive the equations, but we can make some estimates based upon what we already know.

Looking at Figure 12.4, we see that the thickness of the silicon should be smaller than the gate length in order for the gate to control the depletion layer. If the gate length is shorter than the silicon thickness, then the DIBL effects will prevent the MOSFET from turning off. The electric field lines from the gate must be very near the silicon to prevent them from spreading out. This requires that both the top and bottom gate insulator be very thin. A rule of thumb is that:

$$L > 2(t_{Si} + 2t_i) \tag{12.5}$$

where t_{Si} and t_i are the thickness of the silicon layer and the insulator.

As the SOI MOSFET becomes very small, the silicon thickness must become very thin to be able to turn the FET off, and it is highly desired to be able to apply a voltage to both the top and bottom gates. This leads us to the next section on FinFETs.

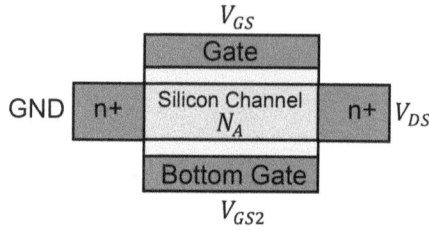

Figure 12.4: Silicon channel in an SOI MOSFET. There is a gate above and below the channel with different gate voltages, a source on the left, and a drain on the right.

12.5 FinFET

The SOI MOSFET works well until the gate length of the MOSFET becomes too small. Then, the MOSFET will not turn off due to DIBL effects! The DIBL performance can be improved if we can apply a voltage both the top and bottom gates. This is called a double-gate MOSFET. Manufacturing a MOSFET that looks like Figure 12.4, with individual control of each bottom gate, is difficult because the electrical connections need to come from the top. A different geometry, called a **FinFET**, can overcome these problems. The FinFET, shown in Figure 12.5, gets its name because it looks like a shark fin sticking up from the substrate.

Electrically, a FinFET is similar to the SOI MOSFET, but the two gates are connected together. Physically, the channel is rotated $90°$ so that the gates are on either side of the silicon channel. To make an electrical connection, the gate is extended to wrap around the top of the fin as well. Thus, the FinFET is really a **tri-gate MOSFET**.

The gate structure of the FinFET provides excellent control over the silicon channel. In fact, if the fin is sufficiently thin, the gate alone controls the potential within the channel, and hence the gate controls the electron and hole concentration within the channel. Therefore, there is no need for the source-substrate and drain-substrate p-n junctions and their depletion regions to prevent current from flow when the MOSFET is off. The depletion region will be created using the gate electrode. This means that the channel region can be undoped, or intrinsic.

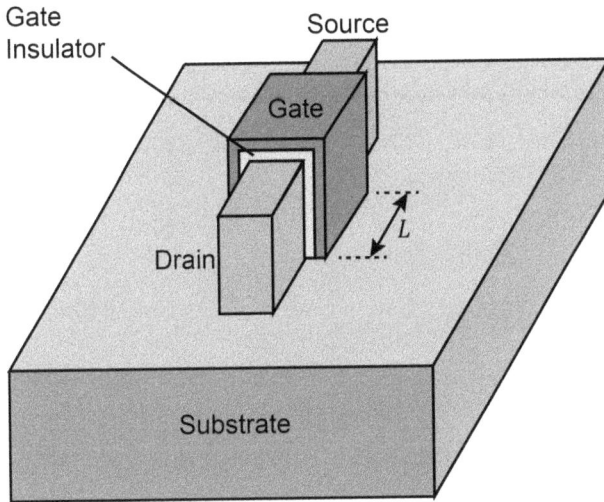

Figure 12.5: FINFET geometry. A "fin" is formed on the substrate. One end of the fin is the drain, and the other end of the fin is the source. The gate wraps around three sides of the fin.

An intrinsic channel provides several advantages. First, the band-to-band tunneling (BTBT) current is greatly reduced, significantly reducing leakage current. Second, electron and hole mobilities are much higher, increasing circuit speed. Third, the channel volume has become so small that the fluctuations in the dopant concentration can influence the threshold voltage. Eliminating the dopants permits a tighter threshold voltage tolerance.

A diagram showing 12 FinFETs on a substrate is shown in Figure 12.6. The fins run horizontally. The gate conductor runs vertically and overlaps multiple fins. To keep the diagram simple, the source and drain were omitted from the diagram, but the source and drain are located in the fin on either side of every gate.

The MOSFET equations from Chapter 5 still apply for the FinFET, but we need to change the width to equal the fin height:

$$W \approx H \tag{12.6}$$

where H is the height of the fin.

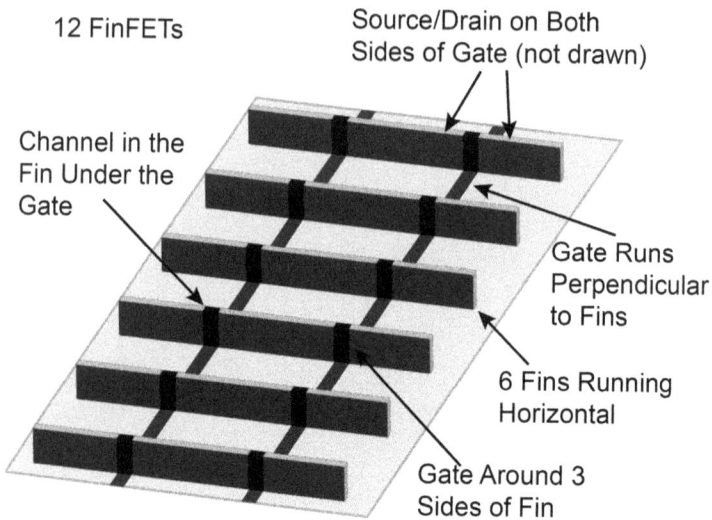

Figure 12.6: Schematic of 12 FinFETs formed using two gate conductors and six fins.

12.6 Gate All Around Transistor

We started with a planar MOSFET in Chapter 5 with a single gate. Then, introduced a dual gate MOSFET in section 12.4 with the SOI MOSFET. The FinFET is really a tri-gate MOSFET. What about a 4-gate MOSFET? I am glad you asked!

Even better gate control, and thus even better DIBL control, is obtained if the gate wraps around all sides of the channel. This is sometimes called a **gate-all-around transistor** because many cross-sectional shapes may be considered. For example, if the cross-sectional area is a thin rectangle, it may be called a nano-ribbon MOSFET. There is also interest in using nanowires, such as carbon nanotubes or silicon nanowires, as the channel.

If the channel dimension becomes smaller than ≈ 5 nm, quantum confinement effects become significant. The energy levels that the electrons may occupy become quantized, much like the atomic orbitals of an atom. This affects things like the threshold voltage of the MOSFET, but the overall operation of the MOSFET will still be similar to the planar MOSFET described in Chapter 5.

12.7 Summary of Key Concepts

- Traditionally transistors shrink by a factor of 0.7 every 2 years.
- With a shrink in transistor geometry, the power density remains the same, the total power consumption drops, the circuit becomes faster, and the cost goes down.
- The source and drain depletion regions interact with each other in short-channel MOSFETs.
- An applied drain voltage lowers the threshold voltage in an effect called drain induced barrier lowering (DIBL).
- Band-to-band tunneling occurs between a p- and n- region, in which electrons tunnel from the valence band of the p- region to the conduction band of the n- region. This is similar to Zener breakdown.
- Many manufacturers have adopted a silicon-on-insulator (SOI) substrate because it has better electrical isolation in a CMOS circuit that traditional CMOS. SOI also provides a speed advantage by reducing the parasitic capacitances.
- Some manufacturers have adopted FinFETs in place of conventional MOSFETs. FinFETs have a section of silicon, called a fin, projecting out of the substrate.
- A channel in a FinFET is surrounded on three sides by the gate, providing superior gate control of the charge in the channel. This reduces the DIBL effect.
- The drain current in a FinFET is better because the channel can be undoped, resulting in a higher carrier mobility.
- The gate all around transistor is the next extension to the multiple-gate MOSFET.

12.8 Problems

1. Consider an NMOS structure with a p-type silicon substrate doped with $N_A = 10^{16}$ cm^{-3} boron atoms. The source and drain are doped n-type with $N_D = 10^{19}$ cm^{-3}. The drawn gate length is $L_{drawn} = 3$ μm. The insulator is SiO$_2$ with a thickness of 30 nm. $N_{ss} = 2 \times 10^{10}$ cm^{-2}. The gate is a polysilicon gate that is heavily doped n-type. The electron

mobility is $\mu_n = 350 \; \frac{cm^2}{V \cdot s}$. $W = 10 \; \mu m$. We will investigate the channel length modulation model in this problem.
   a. What is the flat band voltage?
   b. What is the threshold voltage?
   c. What is the effective gate length, L, with zero applied voltage?
   d. If the drain voltage is $V_{DS} = 3 \; V$, what is the new effective gate length L ?
   Note: This is the same MOSFET as problem #1 in Chapter 11.

2. In a MOSFET, it is called punch-through when the source and drain depletion regions meet (when the effective channel length is zero). Punch-through is a form of non-destructive breakdown for a MOSFET. For the MOSFET from problem #1, what is the drain punch-through voltage?

3. We will see how the DIBL effects impacts this MOSFET. For the MOSFET from problem #1,
   a. Calculate the depletion charge density under the gate using Equation 5.22.
   b. Calculate the depletion charge density under the gate using Equation 12.3 with zero drain voltage.
   c. Calculate the change in threshold voltage due to the shared charge.
   d. Now calculate the change in threshold voltage with $V_{DS} = 3 \; V$.

4. We will make I-V curves for the MOSFET of problem #1. Take into account channel length variation (changes L) and DIBL (changes V_T).

   a. Plot I_{DS} versus V_{DS} for $V_{DS} = 0 \cdots 3 \; V$. Use $V_{GS} = 2V$. Assume that the gate length does not change and there are no DIBL effects.

   b. Now perform the same calculation as part (a), but vary the gate length and the threshold voltage with the drain voltage. Overlay this graph on the graph from part (a).

5. Repeat problems #1c and #1d, but use a gate length of 1.2 µm.

6. Repeat problem #2, but use a gate length of 1.2 µm. The punch-through voltage should be MUCH smaller.

7. Repeat problem #3, but use a gate length of 1.2 µm.

8. In problem #6, when the MOSFET was scaled, the threshold voltage became nearly 0 V for $V_{DS} = 0$ V. Design a threshold implant to adjust the threshold voltage back to its original value. What species (element) and dose should be used? (See Chapter 5.7)

9. It seems that we have solved the DIBL effect problem with a threshold implant, but we haven't. What is the threshold voltage for $V_{DS} = 3$ V? Does the MOSFET turn off?

Problems #1-4 show that we can make a good 3 µm MOSFET. Problems #5-9 show that the same parameters (doping, oxide thickness) cannot be used for a 1.2 µm MOSFET. The MOSFET will not turn off, so it cannot be used for digital circuits. To see why this MOSFET cannot be used for analog circuits either, do problem 10.

10. Using the threshold implant from problem #8, repeat problem #4 using a gate length of 1.2 µm.

The next problem asks you to design a MOSFET. There isn't a unique answer.

11. Design a MOSFET with a gate length of 1.2 µm.

Note: High substrate concentrations cause the mobility to be reduced and band to band tunneling will be important. This is important for very small MOSFETs (gate length < 100 nm) and the design will need to be verified with simulation.

13 Power Transistors

Transistors designed to handle a large amount of power are designed differently than the devices described earlier in this book. The power MOSFET looks substantially different as the current typically flows vertically through the substrate. This book devotes one chapter (this chapter) to power transistors, using the vertical channel MOSFET and the insulated gate bipolar transistor (IGBT) as examples. If you wish to learn more, there are entire books devoted to this topic.

13.1 Power Transistor Requirements

Typical uses of power transistors include controlling motors, AC-to-DC converters, DC-to-AC converters, and DC-to-DC converters. The transistor must have small resistance, high breakdown voltage, low inductance, and low capacitance. The small inductance and small capacitance are required for fast switching speeds and to minimize switching losses.

Table 13.1: Requirements for Power Transistors

Small Resistance
Large Breakdown Voltage
Small Inductance
Small Capacitance

To handle a large current, the series resistance must be very small or the power losses ($P = I^2R$) will be substantial. To handle a large voltage, the depletion regions must be large to prevent avalanche breakdown due to a large electric field.

In power applications, the transistor is used as a switch to control the current through a load (Figure 13.1). The load could be a resistive load, as shown, or some other load such as an electric motor. When the

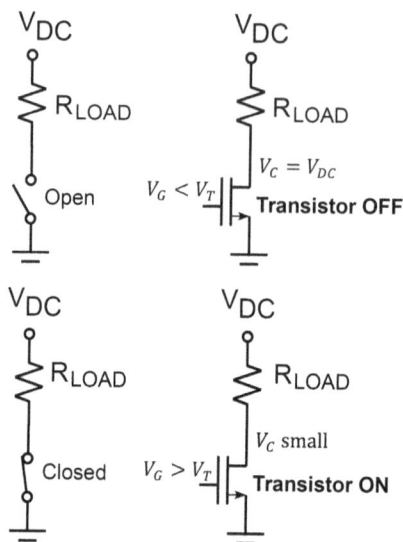

Figure 13.1. Schematic showing how a transistor is used as a switch, using an IGBT as an example. The top row shows the transistor in the open (OFF) position, and the bottom row show the transistor in the closed (ON) position.

transistor is off, the transistor acts like an open circuit and sees the power supply voltage. Thus, the power transistor must be designed to survive a large voltage. When the transistor is on, the transistor should have near zero voltage across it, and the circuit determines the current. The power transistor needs to be designed to have a very small voltage drop to minimize the power losses.

13.2 Power MOSFET

When a MOSFET is used as a switch, there should be very little voltage across it when the MOSFET is turned on. Therefore, the MOSFET should be in the triode mode of operation. When used as a switch, the MOSFET is not operated in saturation because the power losses are too high ($P_{loss} = V_{DC}I_{DS}$). For small drain voltage, the drain current varies nearly linearly with drain voltage. From Table 5.2, and ignoring the V_{DS}^2 term, the drain current is:

$$I_{DS} = \mu_n C_i \frac{W}{L} (V_{GS} - V_T) V_{DS} \tag{13.1}$$

Relating this to Ohm's law, $I = V/R$, we can define a **channel resistance**:

$$R_{ch} = \frac{\partial V_{DS}}{\partial I_{DS}} = \frac{L}{\mu_n C_i W (V_{GS} - V_T)} \tag{13.2}$$

It looks like we could reduce the channel resistance by increasing the gate voltage, but this doesn't work well in practice. The maximum gate voltage is determined by the insulator thickness to prevent breakdown of the gate insulator. The required insulator thickness increases linearly with gate voltage. Thus, ignoring V_T, as the gate voltage increases, t_i increases and C_i decreases, canceling out the benefit of the increased gate voltage.

Example 13.1: (a) Calculate the channel resistance of a MOSFET with a square gate ($W = L$). The insulator is made from SiO_2 and is 10 nm thick, the mobility is 350 cm^2/Vs, and $V_T = 1$ V. The gate voltage is $V_{GS} = 5$ V. (b) Now find the channel resistance for a wide MOSFET with $W = 1000\ L$.

Answer: Find C_i:

$$C_i = \frac{\epsilon_i \epsilon_0}{t_i} = \frac{(3.9)(8.854 \times 10^{-14}\ \text{F/cm})}{10 \times 10^{-7}\ \text{cm}} = 345\ \text{nF/cm}^2$$

(a) Use Equation 13.2 to find the channel resistance (L and W cancel out):

$$R_{ch} = \frac{1}{(350\ \text{cm}^2/\text{Vs})(345\ \text{nF/cm}^2)(5\ \text{V} - 1\ \text{V})} = 2070\ \Omega$$

where we used the relation $1\ s/F = 1\ \Omega$.

(b) Repeating the problem with $W = 1000\ L$, we will get a resistance 1000 × smaller.

$$R_{CH} = 2.07\ \Omega$$

This illustrates why power MOSFETs must have a large width.

Let us look at a MOSFET designed to handle a large voltage. The high voltage is only present when the MOSFET is off. The source, substrate,

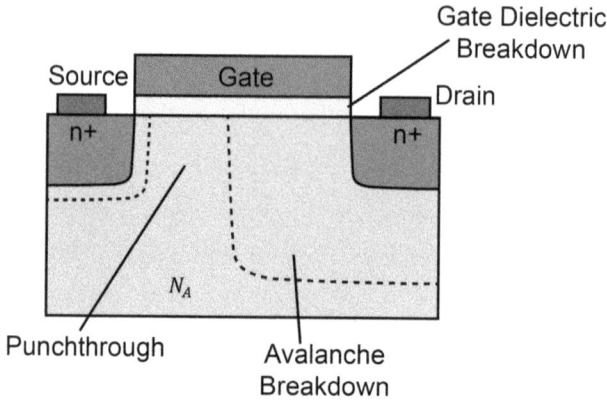

Figure 13.2: A planar MOSFET has three regions where breakdown may occur.

and gate are all at zero volt. ($V_S = V_B = V_G = 0$ V) The drain is at a high voltage. The high voltage causes three concerns, illustrated in Figure 13.2: (1) avalanche breakdown between the drain and substrate, (2) punch-through breakdown if the drain depletion region touches the source depletion region, and (3) the breakdown voltage of the gate insulator. A high avalanche breakdown voltage is achieved by using a lightly doped substrate. That is, N_A is small.

Figure 13.2 shows the depletion region of a typical planar MOSFET with a large drain voltage applied and a small N_A. The depletion region extends far into the substrate because the drain-substrate p-n junction is reverse biased. Remember, the depletion region primarily extends into the lightly doped side of a p-n junction.

A large drain depletion region necessitates a large gate length, L, be used to keep the depletion region from reaching the source to prevent punch-through. Finally, the drain-gate voltage is high, requiring a very thick gate insulator to prevent dielectric breakdown. This combination of a large L and a large t_i (small C_i) results in a very large series resistance for a planar MOSFET designed to handle a large voltage.

The geometry of the power MOSFET can be improved by using a lightly doped drain region, called a drift region, and moving the heavily doped drain to the bottom of the substrate, as shown in Figure 13.3a. This is called a **DMOS**. By using a lightly doped drift region, $N_D^- \ll N_A$, the

DMOS

VMOS

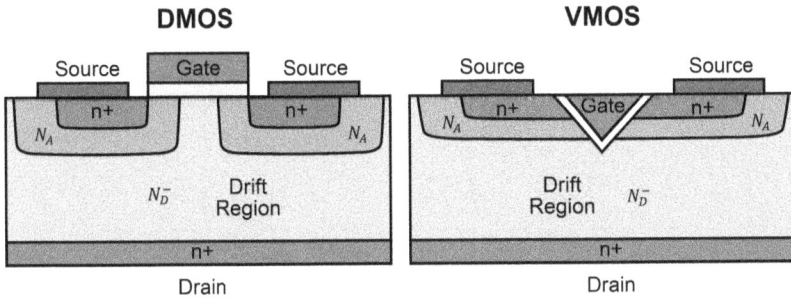

Figure 13.3: Two versions of the power MOSFET.

avalanche breakdown is determined by the doping concentration within the drift region. Furthermore, the depletion region extends into the drift region, and not into the p region. This allows a small gate length to be utilized. Finally, this protects the gate insulator from the high voltage.

It is a win-win, but we moved the high resistance to the drift region. This is the reason for the vertical geometry; to get as large a cross-sectional area as possible to reduce the series resistance of the drift region. The series resistance can be further minimized by using a geometry that rotates the gate, as shown in Figure 13.3b. This is called a **VMOS** due to the v-shaped groove.

The MOSFET equations derived in Chapter 5 still apply. The channel resistance may be found from Equation 13.2. Calculating the series resistance is more difficult because the electrons flow from a single point at the bottom of the v-groove (VMOS) and spread out as they travel through the drift region. As a simple model, let us assume that the current flows through a triangular shape. Then the resistance is approximately twice what we would expect for a rectangular shape:

$$R_{\text{Drift}} = \frac{2\rho L_D^-}{W x_D}$$

(13.3)

where ρ is the resistivity, L_D^- is the thickness of the drift region, x_D is the width of the drain region at the bottom of the drift region, and W is the normal width of the MOSFET.

The large drift region provides ample room for the depletion region, but leads to a high series resistance. The only way to avoid this tradeoff is to make a really wide MOSFET, which not only substantially increases the cost, but it makes the parasitic capacitances very large resulting in large switching power losses. The IGBT, to be introduced in the next section, does a better job when both high voltage and high current are required.

Table 13.2: Typical selection guide for power transistors.

	Low Current	**High Current**
Low Voltage	Planar MOSFET	Planar MOSFET
High Voltage	VMOS	IGBT

13.3 IGBT Principle of Operation

An Insulated Gate Bipolar Transistor (IGBT) is a type of power transistor that takes a vertical channel MOSFET and adds a diode in series. At first glance this may seem like a crazy thing to do, since our goal is to reduce the voltage drop across a transistor, but when the diode is attached to the MOSFET and the n-regions are merged, the diode magically reduces the overall voltage drop across the drift region of the MOSFET.

A cross-section of an IGBT is shown in Figure 13.4. There are three electrodes: the **gate**, the **collector**, and the **emitter**. The names collector and emitter come from the naming scheme of a bipolar junction transistor. This structure is symmetric and may be arranged in a parallel fashion by tiling this layout to make the IGBT carry more current, just like the MOSFET. The collector is at the bottom of the substrate, connected to a heavily doped p+ region, forming a p-n junction between the p+ collector and the n- drift region (called a **base** in the IGBT). The emitter is situated at the top on both sides of the gate. Both portions of the emitter electrode must be tied together. The emitter is connected to both an n+ and a p+ region, where the '+' signs indicate that the doping concentration is large. There are no electrodes connected to the base (drift) region.

Figure 13.4b shows the same cross section, but with two regions identified. These will be analyzed in detail below to understand how

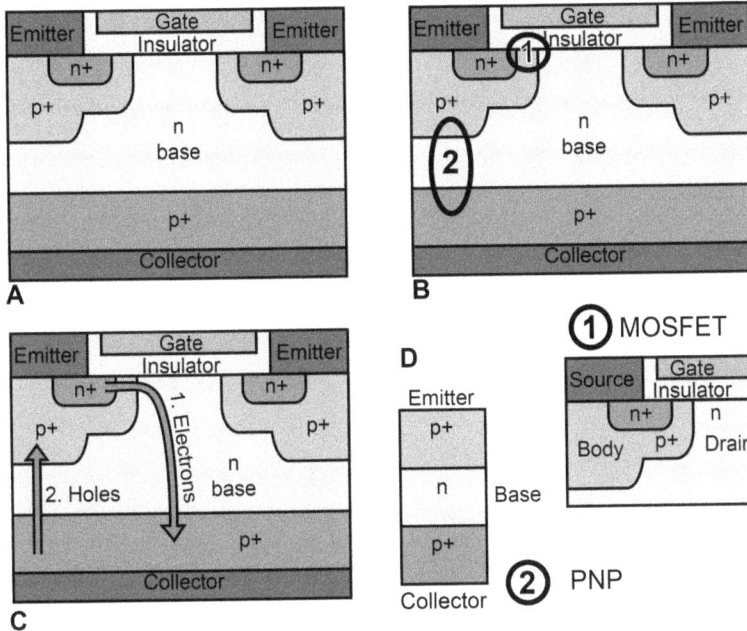

Figure 13.4: (a) Schematic of an IGBT. (b) The location of the MOSFET and PNP are highlighted. (c) The primary paths for electrons and holes are shown. (d) A closeup of the MOSFET and PNP structure.

the IGBT operates. Figure 13.4c shows the same cross section, but with two current paths identified, which correspond with the two areas identified in the upper right. In Figure 13.4d, the detail of regions 1 and 2 are shown. Region 1 corresponds to a MOSFET, and region 2 corresponds to a PNP structure, which has interesting features that will be described later.

In the discussion that follows, we will refer to the different p-n junctions that exist. To simplify the description, we will use the following definitions:

Emitter junction: The p-n junction between the p+ emitter and the n-base.

Collector junction: The p-n junction between the n-base and the p+ collector.

Example 13.2: Consider an IGBT used to control a $1\,\Omega$ load connected to a 100 V power supply. When on, the IGBT has $V_{CE} = 2$ V. How much power is dissipated in the IGBT when the IGBT is (a) off and (b) on. (c) What is the equivalent resistance of the IGBT when on?

Answer: When off, zero current flows so the power dissipated is zero. When on, $I = \dfrac{100\text{ V}}{1\,\Omega} = 100$ A. The power dissipated in the IGBT is:

$$P = IV = 100\text{ A} \times 2\text{ V} = 200\text{ W}$$

This example shows why it is very important to minimize the voltage drop across the IGBT when it is on. The equivalent resistance is

$$R_{eq} = \frac{2\text{ V}}{100\text{ A}} = 0.02\,\Omega$$

This illustrates the need to keep all parasitic resistances in the IGBT to a minimum.

13.3.1 Electrostatics

An IGBT normally has the emitter grounded and a positive voltage is applied to the collector. The gate voltage determines whether the IGBT is on or off. The applied voltages will cause the collector junction to be forward biased, since the collector has the largest positive voltage. Moving towards the emitter, we see that the emitter junction must be reverse biased. Figure 13.5 shows the depletion regions of the IGBT when the IGBT is on ($V_G \geq V_T$) and when the IGBT is off ($V_G \leq V_T$).

Key points:

> The collector junction is always forward biased. It doesn't matter if the IGBT is on or off. The emitter junction is always reverse biased. It doesn't matter if the IGBT is on or off.

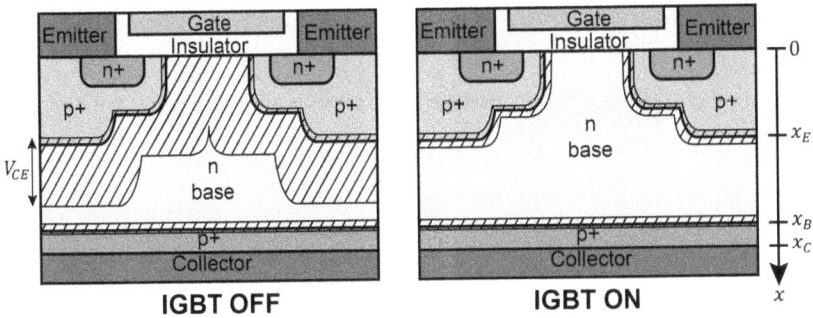

Figure 13.5: Location of the depletion regions for an IGBT that is (a) off and (b) on. The coordinate system, and location of the junctions, is shown on the right.

When the IGBT is off, the collector voltage will be large. This voltage must be dropped across some portion of the device. Since the collector junction is forward biased, and there is no current, this junction must not drop any voltage. The built-in potential will still exist, but the applied voltage across the collector junction is zero. This leaves the emitter junction, which is reverse biased, to support the entire voltage. The emitter is heavily doped, so most of the depletion region extends into the base. The base doping concentration must be low enough to avoid avalanche breakdown (see Chapter 7.7.2), and the length of the base, $x_B - x_E$, must be large enough to prevent punch-through (see Chapter 7.7.3). The base is typically 10's to 100's of μm long. Longer base regions are required for supporting a larger voltage.

When the IGBT is on, the collector voltage should be very small to act as an ideal switch and dissipate very little power. The collector junction is forward biased and a lot of current flows through it. Therefore, the collector junction has a voltage drop around 1 V. The series resistance through the rest of the IGBT, such as the lightly doped base and the channel resistance of the built-in MOSFET, drop another volt. The base behaves in a non-ohmic manner, as will be described later. The reverse biased emitter will actually let current flow due to its coupling with the collector junction, as will also be described later.

13.3.2 Region 1: MOSFET

The portion of Figure 13.5 labeled Region 1 looks like, and acts like, a MOSFET as described in Chapter 5. The MOSFET source, connected to the IGBT Emitter, is grounded, so the equations from Chapter 5 are still valid. The MOSFET drain is connected to the IGBT base and is not heavily doped. The p- region, corresponding to a MOSFET substrate, is also connected to the Emitter and is thus grounded. The threshold voltage is calculated using Equation 5.28. We will find out that the voltage at the surface of the n- region is typically small, but it cannot be zero or no current would flow through the MOSFET. Since V_D is small, the MOSFET is usually in the triode mode when turned on. We can find the voltage drop across the MOSFET portion of an IGBT from the channel resistance:

$$V_{DS} = R_{CH}I_{DS}$$

(13.4)

where R_{CH} is the channel resistance and I_{DS} is the current through the MOSFET.

Since the IGBT is a power transistor, large voltages are typically present. The thickness of the insulating layer determines the breakdown voltage between the gate and emitter. Commercial IGBTs currently use SiO_2 as the insulating layer, which has a breakdown voltage between 1 and 1.5 V/nm. Although the threshold voltage may be small, a larger gate voltage is required to reach the maximum turn-on of the IGBT. The IGBT is generally designed to handle an even larger V_{GE} for robustness because IGBTs are often used for inductive loads where voltage spikes are commonplace. It is therefore desired to keep the insulating layer as thick as possible, but it cannot be too thick as the following example illustrates.

Example 13.3: This example will show why very large gate insulators are not used. What is the threshold voltage for a 1 μm thick insulator of SiO_2 with a p-type polysilicon gate? The substrate is doped with $N_A = 10^{18}$ cm^{-3} and $N_{SS} = 10^{10}$ cm^{-2}.

Answer:

$$\Phi_G = \frac{1}{q}(\chi + E_g) = 4.05\ V + 1.12\ V = 5.17\ V$$

$$E_{iF} = 0.0259\ eV \ln\frac{10^{18}\ cm^{-3}}{10^{10}\ cm^{-3}} = 0.477\ eV$$
$$\text{(Eq. 3.13)}$$

$$\Phi_B = \frac{1}{q}\left(\chi + \frac{E_g}{2} + E_{iF}\right) = 4.05\ V + 0.56\ V + 0.477\ V$$
$$= 5.087\ V$$

$$\Phi_{GB} = \Phi_G - \Phi_B = 5.17\ V - 5.087\ V = 0.083\ V$$

$$C_i = \frac{\epsilon_i \epsilon_0}{t_i} = \frac{(3.9)(8.854 \times 10^{-14}F/cm)}{10^{-4}cm}$$
$$= 3.45\ nF/cm^2$$

$$Q_{SS} = (1.6 \times 10^{-19}C)(10^{10}\ cm^{-2})$$
$$= 1.6\ nC/cm^2$$

$$V_{FB} = \Phi_{GB} - \frac{Q_{SS}}{C_i} = 0.083\ V - \frac{1.6\ nC/cm^2}{3.45\ nF/cm^2} = -0.381\ V$$

$$\phi_{SB} = 2\frac{k_B T}{q}\ln\frac{N_A}{n_i} = 2(0.0259\ V)\ln\frac{10^{18}\ cm^{-3}}{10^{10}\ cm^{-3}} = 0.954\ V$$

$$V_T$$
$$= -0.381\ V + 0.954$$
$$+ \frac{\sqrt{2(11.8)(8.854 \times 10^{-14}F/cm)(1.6 \times 10^{-19}C)(10^{18}cm^{-3})(0.954V)}}{3.45\ nF/cm^2}$$
$$V_T = 164\ V$$

This threshold voltage is too large for most purposes. The problems with this design are (1) the substrate doping concentration is too large, and (2) the insulator thickness is too large. On the positive side, the gate can sustain over 1000 V before breakdown occurs.

13.3.3 Review of Single P-N Junction

The next section will discuss two p-n junctions that are coupled to each other. Before that, let us review a few important facts about p-n junctions. The equation for the current density in a p-n junction is (Equation 7.32):

$$J = qn_i^2 \left(\frac{D_p}{L_p N_D} + \frac{D_n}{L_n N_A} \right) \left[\exp\left(\frac{qV_{app}}{k_B T} \right) - 1 \right] \tag{13.5}$$

This equation may be broken into two pieces: the current density due to electrons and the current density due to holes:

$$J_n = qn_i^2 \left(\frac{D_n}{L_n N_A} \right) \left[\exp\left(\frac{qV_{app}}{k_B T} \right) - 1 \right] \tag{13.6}$$

$$J_p = qn_i^2 \left(\frac{D_p}{L_p N_D} \right) \left[\exp\left(\frac{qV_{app}}{k_B T} \right) - 1 \right] \tag{13.7}$$

Now let us see what percent of the current through the p-n junction is due to the holes. This parameter will be called the **injection efficiency**, for reasons that will make sense when we discuss coupled p-n junctions in the next section. The injection efficiency, γ, is unitless. The equation is:

$$\gamma = \frac{J_p}{J} = \frac{qn_i^2 \left(\frac{D_p}{L_p N_D} \right) \left[\exp\left(\frac{qV_{app}}{k_B T} \right) - 1 \right]}{qn_i^2 \left(\frac{D_p}{L_p N_D} + \frac{D_n}{L_n N_A} \right) \left[\exp\left(\frac{qV_{app}}{k_B T} \right) - 1 \right]} \tag{13.8}$$

Notice that the voltage is going to cancel out. Simplifying, we get:

$$\gamma = \frac{\frac{D_p}{L_p N_D}}{\frac{D_p}{L_p N_D} + \frac{D_n}{L_n N_A}} = \left(\frac{\frac{D_p}{L_p N_D} + \frac{D_n}{L_n N_A}}{\frac{D_p}{L_p N_D}} \right)^{-1} \tag{13.9}$$

Finally, using $L_p = \sqrt{D_p \tau_p}$ and $L_n = \sqrt{D_n \tau_n}$, we get:

$$\gamma = \left(1 + \sqrt{\frac{D_n}{D_p}} \sqrt{\frac{\tau_p}{\tau_n} \frac{N_D}{N_A}} \right)^{-1} \tag{13.10}$$

Example 13.4: What is the injection efficiency of a silicon p-n junction with $N_A = 10^{18}$ cm^{-3} and $N_D = 10^{15}$ cm^{-3}?

Answer: To calculate the injection efficiency, we first need the diffusion coefficients, which requires the electron and hole mobilities. Keep in mind that when working with p-n junctions, we are always working with minority carriers. Therefore, μ_n and D_n are calculated from the doping concentration on the p- region, where the electrons are the minority carriers. Similarly, μ_p and D_p are calculated from the doping concentration on the n- region, where the holes are the minority carriers. From Table B.3, the electron and hole mobilities are:

$$\mu_n = 261 \; \frac{\text{cm}^2}{\text{V} \cdot \text{s}}$$

$$\mu_p = 458 \; \frac{\text{cm}^2}{\text{V} \cdot \text{s}}$$

The diffusion coefficients are calculated as follows:

$$D_n = \frac{k_B T}{q} \mu_n = 0.0259 \text{ V} \times 261 \; \frac{\text{cm}^2}{\text{V} \cdot \text{s}} = 6.76 \; \frac{\text{cm}^2}{\text{s}}$$

$$D_p = \frac{k_B T}{q} \mu_p = 0.0259 \text{ V} \times 458 \; \frac{\text{cm}^2}{\text{V} \cdot \text{s}} = 11.9 \; \frac{\text{cm}^2}{\text{s}}$$

The minority carrier lifetimes on the n- and p- region are found from Table C.1:

$$\tau_n = 2.82 \times 10^{-7} \text{ s}$$

$$\tau_p = 1.28 \times 10^{-3} \text{ s}$$

We can now find the injection efficiency. The units are not shown to save space because the same term is always in the numerator and denominator, and therefore the units will all cancel.

$$\gamma = \left(1 + \sqrt{\frac{6.76}{11.9}} \sqrt{\frac{1.28 \times 10^{-3}}{2.82 \times 10^{-7}} \frac{10^{15}}{10^{18}}} \right)^{-1} = 0.9517$$

The injection efficiency is a value between 0 and 1. For example, if the injection efficiency is 0.9, then 10% of the collector current is carried by electrons and 90% of the collector current is carried by holes. If we want the current to be dominated by the holes, then we want to make N_D as small as possible and N_A as large as possible. The other terms all have a square root, so their impact is smaller than that of the doping concentrations.

Before moving on to the coupled p-n junctions, it is helpful to review one more concept: the excess hole concentration in the neutral n-region. The concepts of the long diode and the short diode are very important here, and it may be helpful to re-read section 7.6. Consider a long diode, in which the holes are free to diffuse as far as they wish before recombining. The excess hole concentration is shown in Figure 7.10. The holes enter on the left side because of the forward biased p-n junction. The holes diffuse to the right, with the hole concentration decreasing as the holes recombine.

The excess hole concentration for a short diode is also shown in Figure 7.10. The holes enter on the left side due to the forward biased p-n junction, just like in the long diode. However, because the width of the n-region is physically small, the excess hole concentration is nearly zero a short distance from the p-n junction. We can either calculate the excess hole concentration exactly using the continuity equation, or we may use the approximation that the holes reach the right end before recombining due to the short distance involved. If recombination is negligible, then the excess hole concentration is a nearly straight line, as shown in Figure 7.10.

The important difference between the long diode and short diode is that in a long diode, all the holes recombine before reaching the right edge. In a short diode with negligible recombination, all the holes reach the right edge. Don't let the fact that the excess hole concentration is zero at the right edge deceive you. The diffusion rate is constant because the slope is constant. Therefore, the rate at which holes move to the right is constant in a short diode. That is, in a short diode all the holes that enter the n- region on the left also leave the n- region on the right. In the long diode, the exponential decay means that the slope becomes shallower towards the right edge, indicating that the hole diffusion current is becoming smaller. In a long diode, none of the holes reach the right edge.

We are now ready to look at the interaction of two p-n junctions that are near each other.

13.3.4 Region 2: Coupled P-N Junctions

The collector junction consists of a p-n junction between the collector and base. The emitter junction consists of a p-n junction between the emitter and base. Because these two junctions share a common n-region, the base, this arrangement is commonly called a **PNP** region.

The base region is doped n-type. This also the drain of the MOSFET. The electrons that flow from the source into the drain move through the base to the collector. The electrons cannot flow through the emitter junction because the emitter junction is reverse biased. All of the electrons move toward the forward biased collector junction. The electron current in the collector junction is then:

$$I_{C,n} = I_{DS} \tag{13.11}$$

where $I_{C,n}$ represents the electron component of the collector junction current. To get a large total current through the IGBT, we would like to get a lot of holes to contribute to the collector current as well as the electrons.

To find the total collector current, we will make use of a parameter called the injection efficiency, defined in the last section. The injection efficiency is the ratio of the hole current to the total current through the collector junction. The injection efficiency repeated here, but with labels for the Base and Collector portions of the p-n junction:

$$\gamma = \left(1 + \sqrt{\frac{D_{nC}}{D_{pB}}} \sqrt{\frac{\tau_{pB}}{\tau_{nC}}} \frac{N_{DB}}{N_{AC}} \right)^{-1} \tag{13.12}$$

Since the applied voltage cancels out, the injection efficiency is a fixed number, and it must be between 0 and 1. The collector hole current may be written as:

$$I_{C,p} = \gamma I_C \tag{13.13}$$

To find the total collector current in terms of the drain current, we will start by rewriting the collector hole current:

$$I_{C,p} = \gamma I_C = \gamma(I_{C,p} + I_{C,n}) = \gamma I_{C,p} + \gamma I_{C,n} \tag{13.14}$$

Solving for $I_{C,p}$:

$$I_{C,p} = \frac{\gamma}{1 - \gamma} I_{DS} \tag{13.15}$$

Then the collector current is found to be:

$$I_C = \frac{I_{C,p}}{\gamma} = \frac{1}{1 - \gamma} I_{DS} \tag{13.16}$$

That is, the drain current is amplified by a factor:

$$\beta = \frac{I_C}{I_{DS}} = \frac{1}{1 - \gamma} \tag{13.17}$$

where β is the current gain. The current gain is made large by making the injection efficiency large. This is accomplished by making the doping ratio between the collector and base as large as possible. That is, the collector p+ region should be heavily doped, and the base region should be lightly doped. A lightly doped base is compatible with the desire to prevent avalanche breakdown.

Unfortunately, we will never achieve a large current gain ($\beta \gg 1$) except at low collector current density. The problem is that the hole concentration in the base will become very large, requiring the electron concentration to increase to maintain charge neutrality. This is called high-level injection and causes β to decrease. Since we need high-level injection (see section on Base resistance), β is closer to 1, and the PNP structure and MOSFET carry nearly the same current.

So now we get to the last step of the PNP operation, and this is where the magic happens. We have found that there are a lot of holes being injected into the base from the collector region. Where do the holes go? The answer is two places. First, a few of the holes recombine with electrons that don't make it to the collector. Hopefully the recombination is a negligible effect. Second, most importantly, the holes diffuse to the emitter junction. This transport is by diffusion, and not due to an electric field, and thus **ideally there is no voltage drop across the base region**. The holes enter the depletion region of the emitter junction, and the built-in electric field within the reverse biased junction sweeps the holes to the emitter p+ region. In fact, this is necessary to set up the gradient in the hole concentration to obtain a diffusion current.

Figure 13.6 shows the electric field within the two depletion regions and the excess carrier concentrations between the depletion regions. The emitter junction is reverse biased and therefore the excess carrier concentrations are negative adjacent to the emitter depletion region. The collector junction is forward biased and therefore the excess carrier concentrations are positive and large next to the collector depletion region. The excess hole concentration in the base region must have a large positive slope, causing the holes to diffuse to the left. If the diffusion current is large enough, then no electric field is required for the hole transport across the base and there is no voltage drop across the n-base region.

13.3.5 Base Resistance

The collector diode provides a second feature. The resistance in the base is determined by both the electron and hole concentrations. Without current flow, the base resistance would be dominated by the lightly doped base region. But with current flow, with a lot of holes flowing from the collector diode (assuming a high injection efficiency), the overall carrier concentration is generally much greater than the doping concentration. This is called **high level injection**. The resistivity is determined by the electron and hole concentration, and not by the doping concentration in high-level injection.

In high level injection, the minority carrier concentration becomes larger than the doping concentration. The electron concentration must equal the hole concentration: $n \approx p$ (see section 7.8). A great benefit of this is that the resistance of this region is greatly reduced. Thus, the resistance of a thick base region can be small.

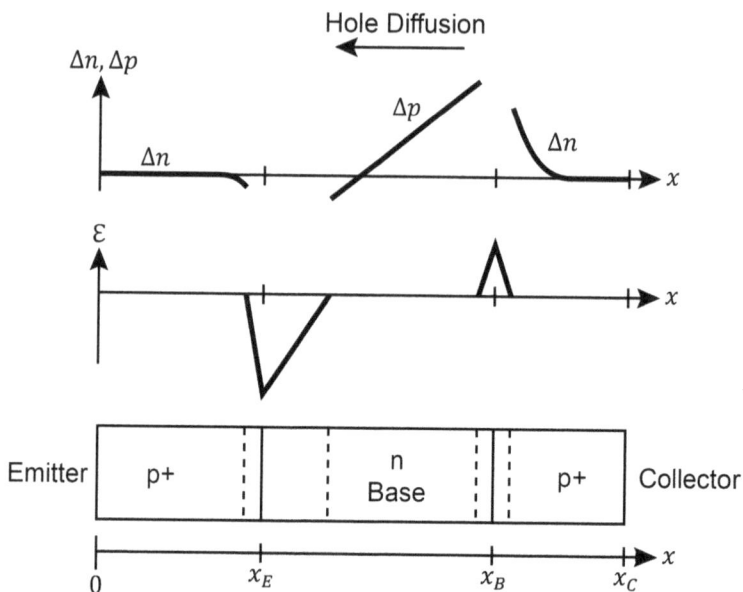

Figure 13.6: These plots show the excess carrier concentration and electric field within the PNP region.

Thus, we can make the argument that the base resistance is small because the hole current is dominated by diffusion current, which has no voltage drop, and the base region is in high level injection, which reduces the base resistance when high levels of current is flowing.

13.3.6 IGBT Current

Calculating the current through an IGBT is difficult because we need the current through the MOSFET. If we can find the current through the MOSFET, then we can find the collector current from Equation 13.16. If we ignore any voltage drop across the base (ideal case), then we can relate the collector current, voltage drop across the collector junction, and V_{DS} by:

$$V_{DS} = V_C - V_{pn}$$

(13.18)

Using the equation for I_{DS} from Chapter 5, the equation for current in a diode (Equation (7.32)), and using Equations (13.16) and (13.18), we can solve for the collector current. These equations cannot be solved by hand and the solution must be found numerically.

Let us try to understand the current through an IGBT step by step. First, in Figure 13.7a, we see the current when the IGBT is off. The current is zero until breakdown occurs.

Second, for a very small collector voltage, the current is limited by the diode. There isn't much current until we get to approximately 1 V. Around 1V, the current starts to increase very rapidly. This is shown in Figure 13.7b. In this region the current may be calculated using the equations in Chapter 7 as if the IGBT were simply a diode.

Third, once the diode is fully turned on, the current is limited by the MOSFET. For illustrative purposes, the MOSFET I-V curves are placed on the IGBT I-V curve in Figure 13.7c. The drain voltage is close to zero when $V_C \approx 1$ V, so the MOSFET curves start at 1 V. Although it is difficult to get an exact answer for the collector current (see above), we can get an estimate by calculating the MOSFET current using $V_{DS} = V_C - 1$ V. The collector current is then βI_{DS}.

Finally, we put all the pieces together and get the IGBT I-V curve shown in Figure 13.7d. There are three regions numbered 1-3 in the figure. When the IGBT is off, the circuit should operate in region 1. The circuit should be designed so that the IGBT is operating in region 2 when the IGBT is on. This puts the MOSFET in the triode regime with a very small V_{DS} and small V_{CE}.

The IGBT should not be operated in region 3, where the MOSFET is in saturation. There is a large voltage and large current, resulting in large power losses. The only reason for operating in region 3 is when the IGBT is turning on or off and the IGBT must transition through this region for a brief amount of time.

13.4 Summary of IGBT Operation

To summarize, application of a gate voltage turns on the MOSFET. The MOSFET sends electrons into the base region, which in turn flow into the collector junction. The electron current forces a hole current to flow, causing holes to flow into the base region. The base region is now flooded with both electrons and holes. The holes diffuse to the emitter junction, where they are collected by the electric field. The fact that the hole current is due to diffusion means that they are not limited by

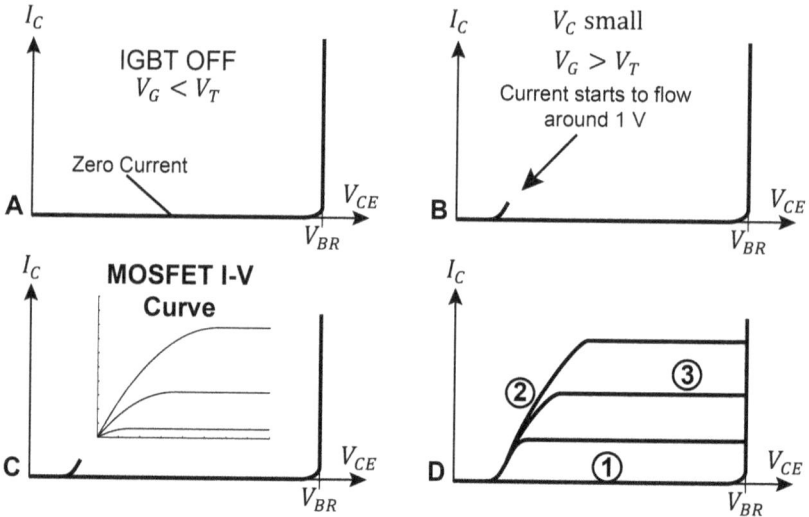

Figure 13.7: IGBT Current. (a) IGBT is turned off. (b) A small collector voltage is applied; looks like a diode. (c) The center region looks like a MOSFET I-V curve. (d) Complete IGBT I-V Curve.

resistivity in the classical sense, permitting a higher current density to flow without a large voltage drop.

If the gate voltage is below the threshold voltage, the IGBT will be off. The collector junction is forward biased and the emitter junction is reverse biased. The voltage is supported by the emitter junction, but primarily in the lightly doped base region. The base region must be lightly doped to prevent avalanche breakdown and long to prevent punch-through.

13.5 IGBT Switching

IGBTs do not turn on or off very fast compared to a MOSFET. Having the holes diffuse through the base region is great for obtaining a high current density, but it causes problems when trying to turn off the IGBT. When the gate voltage drops below the threshold voltage, the MOSFET turns off and the flow of electrons into the base stops. The electrons have to move from the top of the base region to the collector junction, which takes some time. Then, the electrons finish moving through the collector junction. During this time, holes are still being injected into the base.

Once the flow of holes injected from the collector is stopped, the holes must deplete from the base region. This takes some time because they must diffuse to the emitter junction, or they must recombine. Diffusion is a slower process than drift by an electric field. IGBTs are substantially slower than a power MOSFET.

During the turn-off transient, current continues to flow through the IGBT for a significant amount of time, although at a much lower magnitude.

13.6 Latch-Up Prevention

Early IGBTs had a problem wherein they would easily latch-up in the on position and could not be turned off. This occurs because of a parasitic NPN configuration consisting of the n+ on the emitter, the p+ of the MOSFET substrate, and the n-base region. The base-p+ emitter junction wants to be forward biased because the voltage from the collector drops to the base, then the MOSFET substrate (p+), before reaching the ground at the emitter n+. A forward biased n+ to substrate junction will inject electrons into the p-region, which are then swept into the base by the reverse biased emitter junction. These electrons into the base cannot be turned off by the gate because they are bypassing the MOSFET. Thus, the IGBT cannot be turned off if this occurs. To prevent latch-up, the IGBT must be designed with the n+ region and the p+ region at the emitter shorted together so that this junction cannot become forward biased.

13.7 Summary of Key Concepts

- In a power MOSFET, the drain current runs vertically through the substrate. The drain contact is on the backside of the wafer.
- In a power MOSFET, the source and body are tied together.
- In a power MOSFET, the gate current will often run in a vertical direction.
- SiC can be used as a material for power MOSFETs due to its critical electric field (i.e., it has a higher breakdown voltage for the same doping concentrations).
- The lightly doped drift region permits a high breakdown voltage.
- The lightly doped drift region results in a high series resistance.

- The depletion region in the drift region keeps the drain voltage from being near the gate electrode, preventing dielectric breakdown when the MOSFET is off.
- The IGBT is a power transistor capable of handling both a high voltage and a high current.
- IGBTs handle a higher power than MOSFETs, but are slower than MOSFETs.
- The current flows through the thickness of the substrate.
- Internal to the IGBT is a MOSFET that controls the current to the base region, turning the IGBT on and off.
- When electron current passes through the collector junction, a larger hole current will be created.
- Holes diffuse through the base region, from the collector junction to the emitter junction.
- A thick base region is required to support a large voltage in the off state.
- The emitter must short the p+ and n+ regions to prevent latch-up.

13.8 Problems

1. Draw a cross-section of a VMOS PMOS power transistor, similar to Figure 13.3.

2. A UMOS is a power transistor where the gate runs vertically, instead of at an angle as in a VMOS. Draw a schematic of an n-channel UMOS.

3. Consider a DMOS or VMOS with that is off ($V_{GS} < V_T$), but there is a large drain voltage. Draw a schematic of the DMOS or VMOS and indicate the locations and relative size of the depletion regions.

4. Consider a power MOSFET made from silicon that needs to have a drain breakdown voltage greater than 1000 V. Use $N_A = 10^{17}$ cm^{-3}.
   a. What is the maximum doping concentration in the drift region? (Consider avalanche breakdown)
   b. What is the minimum thickness of the drift region? (Consider punch-through)

5. Repeat problem #4 for a SiC MOSFET.

6. Compare your answers from problem #4 and #5.
   a. By what factor does the resistance decrease due to the increase in doping concentration when using SiC versus Si (compare parts a)?
   b. By what factor does the resistance decrease due to the decrease in thickness of the drift region when using SiC versus Si (compare parts b)?
   c. What is the overall decrease in resistance when using SiC versus Si?

   Note: This improvement in resistance does not involve a change in geometry, so the capacitance is approximately the same. This makes SiC very attractive for power devices.

7. When a power transistor is made, it is typically arrayed many times in parallel to increase current flow. Let's say that there are 50 cells, where each cell is half the schematic shown in Figure 13.3. Use $W = 1$ mm, $x_D = 20$ μm, $V_T = 1$ V, an SiO_2 gate insulator with $t_i = 20$ nm, and the channel mobility is $\mu_n = 300 \frac{cm^2}{Vs}$. (Total MOSFET size: 1 mm x 1 mm)
   a. Using $V_{GS} = 10$ V, what is the channel resistance of a single cell?
   b. What is the equivalent channel resistance of 50 cells?
   c. Using your results from problem #3, what is the drift region resistance?
   d. What is the drift region resistance for 50 cells?
   e. What is the total resistance, considering all 50 cells?
   f. Which dominates the overall resistance, the channel resistance or the drift resistance?

8. Repeat problem #7 for a SiC MOSFET. Use a channel mobility of $\mu_n = 30 \frac{cm^2}{Vs}$. Additional questions:
   a. Compare your results for part f between a Si and SiC MOSFET. What do you notice?
   b. Which device has a lower overall resistance? The Si or the SiC MOSFET?

9. Draw a cross-section of a p-channel IGBT, similar to Figure 13.4. Indicate the path for electrons and holes.

10. The MOSFET within the IGBT does not need to be planar. It could be a VMOS, or even run vertically (called a UMOS). Draw a schematic of an IGBT that uses a VMOS.

11. Consider an n-channel IGBT with the following parameters:
    Collector: $N_A = 10^{19}$ cm^{-3}
    Base: $N_D = 10^{16}$ cm^{-3}
    Emitter: $N_A = 10^{17}$ cm^{-3}
    Drain: $N_D = 10^{19}$ cm^{-3}
    Gate insulator: SiO$_2$, $t_i = 25$ nm
    Effective mobility, $\acute{\mu}_n = 300 \frac{cm^2}{V \cdot s}$
    N_{SS}: 10^{10} cm^{-2}
    P+ poly gate
    Gate length: 1 μm
    Cell width = 50 μm, Gate width of each cell: 5 mm, 100 cells
    Collector Area: 0.25 cm2
    Effective gate width: 500 mm
    a. What is the threshold voltage?
    b. What is the injection efficiency?
    c. What is the current gain of the PNP transistor, β?

12. The IGBT described in problem #11 is used in a circuit with 10 A is flowing through it when the IGBT is on with $V_{GS} = 10$ V.
    a. What is the voltage drop across the collector junction?
    b. How much current is flowing through the MOSFET?
    c. Assuming that the MOSFET is in the triode region, what is V_{DS}?
    d. What is the voltage drop V_C across the IGBT?

    Note: With this large current density, β is much smaller than we calculated.

13. Consider the IGBT described in problem #11.
    a. What is the avalanche breakdown voltage?
    b. How thick does the base region, between the collector and emitter, need to be to prevent punch-through before avalanche breakdown?

14. One of the goals of IGBT design is to put the base into high-level injection. This is both a good thing (reduces base resistance) and bad thing (reduces β). Consider the IGBT from problem #11 and #12.
    a. What is Δp at the collector junction, assuming 10 A of collector current?
    b. Is the hole concentration larger than the base doping concentration, putting the base into high-level injection?
    c. Let us recalculate β. The base doping concentration N_{DB} initially set the electron concentration n. In high level injection $n \approx p$. Therefore, use Δp in place of N_{DB} and recalculate β.

Appendix A: Material Properties

A.1 Select Semiconductor Properties at 300 K

	Silicon	Ge	GaAs	GaN	SiC (4H)
Band gap (eV)	1.12	0.67	1.43	3.40	3.26
Conduction band effective density of states, N_C (cm^{-3})	3.51×10^{19}	1.02×10^{19}	4.35×10^{17}	2.30×10^{18}	1.23×10^{19}
Valence band effective density of states, N_V (cm^{-3})	1.87×10^{19}	5.64×10^{18}	7.57×10^{18}	1.80×10^{19}	4.58×10^{18}
Intrinsic carrier concentration, n_i (cm^{-3})	10^{10}	1.79×10^{13}	1.77×10^{6}	1.77×10^{-10}	3.10×10^{-9}
Relative Dielectric constant, ϵ_S	11.8	16.0	13.1	9.5	9.7
Electron Affinity, χ (eV)	4.05	4.0	4.07	3.5	3.7
Thermal conductivity, κ (W/m·K)	150	60	55	195	370

Interesting notes:

1. Diamond has a thermal conductivity of 2200 (W/m·K), which is extremely large.
2. SiO$_2$ has a band gap of 9.1 eV, making it a fantastic insulator with a very high breakdown voltage.

3. The high dielectric constant of these semiconductors causes them to reflect a lot of light, producing a mirror-like surface. This is not a good feature for solar cells, which are supposed to absorb light.

4. Materials with a large band gap, such as SiC and GaN, are transparent. The energy of a photon of visible light is not sufficient to move an electron from the valence band to the conduction band, so visible light cannot be absorbed.

A.2 Select Metal Work functions

Symbol	Name	Work function
Al	Aluminum	4.1 V
Au	Gold	5.2 V
Cr	Chromium	4.5 V
Mo	Molybdenum	4.5 V
Ni	Nickel	5.1 V
Ti	Titanium	4.33 V
W	Tungsten	4.7 V

The reported values for the work function of the metals can vary a lot in the literature. For example, the work function of W is reported to be between 4.3 and 5.2 V.

When measuring the work function of a material, the observed value depends critically on the quality of the surface from which electrons are extracted. If there is any contamination, or chemical bonds, the value will differ. This means that aluminum in free space may have one work function, but aluminum on SiO_2 may have a different work function, aluminum on silicon may have a third work function, and aluminum on SiC may have a fourth work function. Finally, the work function may vary as the aluminum is annealed. The conclusion to take from this variability is that the work functions reported in the table are a great starting point, but the work function for an actual device may differ slightly.

A.3 Select Dielectric Material Properties

The following table lists some insulators that are candidates for use as a gate insulator. SiO_2 and HfO_2 are the most commonly used materials.

Material	Relative Dielectric Constant	Band Gap Energy	Critical Electric Field
SiO_2	3.9	9.1 eV	1.5 V / nm
Si_3N_4	7	5.3 eV	
Al_2O_3	9	8.8 eV	
HfO_2	25	5.8 eV	

Appendix B: Electron and Hole Mobility and Drift Velocity

Values for the electron and hole drift velocity, and their mobilities, are reported in this appendix. These values are for carriers that are far from the surface of a semiconductor, and hence not affected by surface defects. These are called bulk mobilities and bulk drift velocities. The drift velocities and mobilities at the surface of a semiconductor are significantly smaller than the values reported in this Appendix.

B.1 Pure (intrinsic) semiconductors

Semiconductor (Intrinsic)	Electron mobility, $\mu_{n,max}$ $\left(\dfrac{cm^2}{V \cdot s}\right)$	Hole mobility, $\mu_{p,max}$ $\left(\dfrac{cm^2}{V \cdot s}\right)$
Germanium	3900	1900
Silicon	1330	460
Diamond	2800	2000
GaAs	9280	486
GaN	1600	175
SiC	900	117

Note: Do not use these mobility values when doing homework problems. Always calculate the mobility using the tables and graphs in the next section.

B.2 Doped (extrinsic) Semiconductor Mobility Model

In general, the mobility as a function of doping concentration may be calculated from the following model [1]:

$$\mu(N) = \mu_{min} + \frac{\mu_{max} - \mu_{min}}{1 + \left(\dfrac{N}{N_{ref}}\right)^{\gamma}} \tag{B.1}$$

The value of μ_{max} is the mobility for a pure semiconductor. The constants are found from the following table. N is the total doping concentration, regardless of whether the dopants are acceptor atoms or donor atoms. That is, $N = N_A + N_D$.

Semiconductor		μ_{min}	μ_{max}	N_{ref}	γ
Silicon [2]	μ_n	88.3	1330.3	1.295×10^{17}	0.891
	μ_p	54.3	461.2	2.35×10^{17}	0.88
GaN [3]	μ_n	100	1600	3×10^{17}	0.7
	μ_p	10	175	2.5×10^{17}	1.5
SiC [4]	μ_n	35	900	2×10^{17}	0.72
	μ_p	33	117	1×10^{19}	0.5
GaAs [5]	μ_n	500	9400	6×10^{16}	0.394
	μ_p	20	491.5	1.48×10^{17}	0.38
Germanium [6]	μ_n	0	3900	1×10^{17}	0.5
	μ_p	0	1900	7×10^{16}	0.6

The electron and hole mobility as a function of doping concentration are shown in Figures B.1 and B.2. Notice that the mobility is the same whether the wafer is doped with acceptor atoms or donor atoms. The mobility is impacted by any impurity in the semiconductor. The mobility does not change much at low doping concentrations, but changes significantly at higher doping concentrations.

Table B.3: Electron mobility and hole mobility for silicon and GaN for select doping concentrations.

$N_A + N_D$ (cm^{-3})	Silicon		GaN	
	$\mu_n \left(\dfrac{cm^2}{V \cdot s}\right)$	$\mu_p \left(\dfrac{cm^2}{V \cdot s}\right)$	$\mu_n \left(\dfrac{cm^2}{V \cdot s}\right)$	$\mu_p \left(\dfrac{cm^2}{V \cdot s}\right)$
10^{14}	1328	461	1595	175
10^{15}	1314	458	1573	175
10^{16}	1215	437	1473	174
10^{17}	781	331	1125	142
10^{18}	261	143	551	28
10^{19}	114	69	219	11

Table B.4: Electron mobility and hole mobility for SiC and GaAs for select doping concentrations.

$N_A + N_D$ (cm^{-3})	SiC		GaAs	
	$\mu_n \left(\dfrac{cm^2}{V \cdot s}\right)$	$\mu_p \left(\dfrac{cm^2}{V \cdot s}\right)$	$\mu_n \left(\dfrac{cm^2}{V \cdot s}\right)$	$\mu_p \left(\dfrac{cm^2}{V \cdot s}\right)$
10^{14}	896	117	8704	464
10^{15}	881	116	7847	430
10^{16}	810	114	6310	367
10^{17}	573	109	4256	273
10^{18}	242	97	2370	174
10^{19}	84	75	1149	99

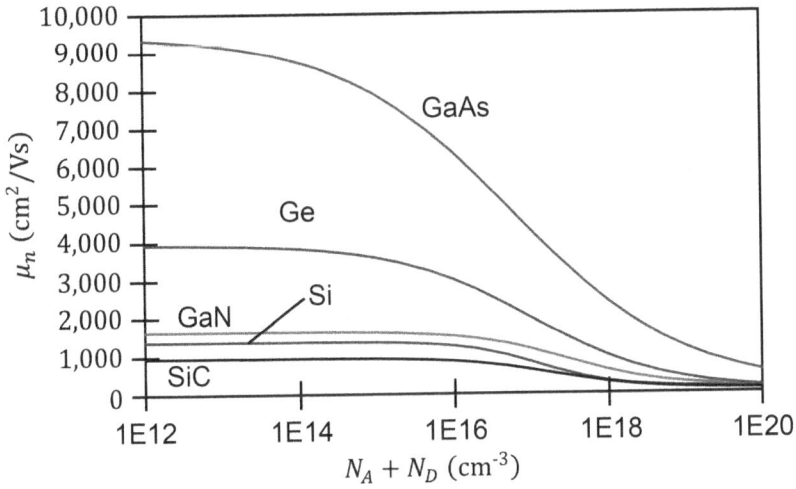

Figure B.1: Electron mobility for five semiconductors. From top to bottom: GaAs, Ge, GaN, Si, and SiC.

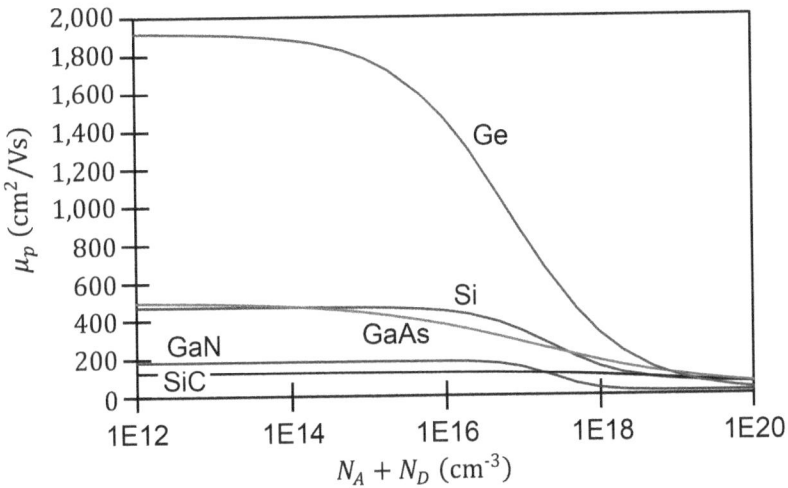

Figure B.2: Hole mobility for five semiconductors.

B.3 Velocity Saturation

The electron and hole velocity as a function of electric field is shown in Figures B.3 and B.4. These are typical values and will vary with doping concentration. The maximum drift velocity, called the saturation velocity, is shown in Table B.5 for a select number of materials.

Table B.5: Saturation velocity for a number of semiconductors.

Material	$v_{sat,n}$ ($\times 10^7$ cm/s)	$v_{sat,p}$ ($\times 10^7$ cm/s)
Silicon	1.05	0.94
4H-SiC [7]	1.83	0.86
GaN	2.8	
GaAs	2.0	0.9
InP	2.5	0.7
Diamond	1.0	1.1
Germanium	0.65	0.63

In this book we have assumed that the drift velocity increases linearly with electric field until the drift velocity saturates. This is a good approximation for many semiconductors, such as silicon, as shown in Figure B.3. However, this is not accurate for all semiconductors.

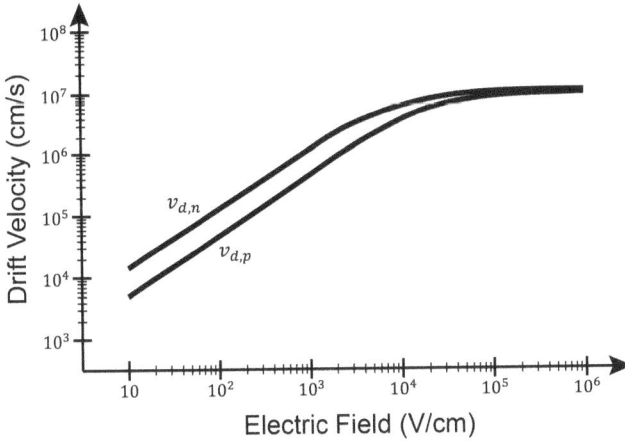

Figure B.3: Electron and hole drift velocity in lightly doped silicon as a function of the electric field.

Appendix C: Electron and Hole Lifetime Data

The lifetime of an electron or hole depends upon the doping concentration, number of defects in the semiconductor, and number of excess carriers in the semiconductor. This appendix provides numbers for the lifetime of the semiconductors under ideal conditions: very few defects and low excess carrier concentration.

C.1 Silicon

Equations have been reported in the literature for "good" values of the lifetime of electrons and holes in silicon. Technically, this is the minority carrier lifetime and may be used in SRH recombination calculations (not covered in this book) for refinement [8].

$$\tau = \frac{1}{3.45 \times 10^{-12} N_A + 9.5 \times 10^{-32} N_A^2}$$

$$\tau = \frac{1}{7.8 \times 10^{-13} N_D + 1.8 \times 10^{-31} N_D^2}$$

The doping concentration (N_A and N_D) must be in units of cm^{-3}, and the lifetime has units of seconds. Use the top equation if the silicon is doped with acceptor atoms, and the bottom equation if the silicon is doped with donor atoms.

Table C.1: Carrier lifetime for silicon for select doping concentrations.

N_A or N_D (cm^{-3})	τ (s) (doped with N_A)	τ (s) (doped with N_D)
10^{14}	2.90×10^{-3}	1.28×10^{-2}
10^{15}	2.90×10^{-4}	1.28×10^{-3}
10^{16}	2.90×10^{-5}	1.28×10^{-4}
10^{17}	2.89×10^{-6}	1.25×10^{-5}
10^{18}	2.82×10^{-7}	1.04×10^{-6}
10^{19}	2.27×10^{-8}	3.88×10^{-8}

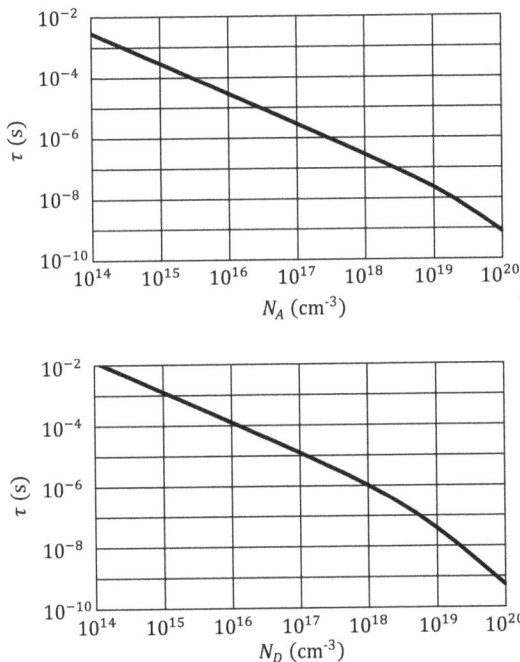

Figure C.1. The carrier lifetime for silicon as a function of doping concentration. If the silicon is doped p-type, use the top graph. If the silicon is doped n-type, use the bottom graph.

References

[1] D. M. Caughey and R. E. Thomas, "Carrier mobilities in silicon empirically related to doping and field," *Proceedings of the IEEE,* vol. 55, pp. 2192-2193, 1967.

[2] N. Arora, J. R. Hauser, and D. J. Roulston, "Electron and Hole Mobilities in Silicon as a Function of Concentration and Temperature," *IEEE Trans. Electron Devices,* vol. 29, pp. 292-295, 1982.

[3] S. Vitanov, "Simulation of High Electron Mobility Transistors," PhD, Institute for Microelectronics at Technische Universität Wien, 2010.

[4] T. T. Mnatsakanov, M. E. Levinshtein, L. I. Pomortseva, and S. N. Yurkov, "Carrier mobility model for simulation of SiC-based electronic devices," *Semiconductor Science and Technology,* vol. 17, pp. 974-977, 2002.

[5] M. Sotoodeh, A. H. Khalid, and A. A. Rezazadeh, "Empirical low-field mobility model for III–V compounds applicable in device simulation codes," *Journal of Applied Physics,* vol. 87, pp. 2890-2900, 2000.

[6] M. B. Prince, "Drift Mobilities in Semiconductors. I. Germanium," *Physical Review,* vol. 92, pp. 681-687, 11/01/1953.

[7] J. H. Zhao, V. Gruzinskis, Y. Luo, M. Weiner, M. Pan, P. Shiktorov*, et al.*, "Monte Carlo simulation of 4H-SiC IMPATT diodes," *Semiconductor Science and Technology,* vol. 15, p. 1093, 2000.

[8] Y. Taur and T. H. Ning, *Fundamentals of Modern VLSI Devices,* 2nd ed.: Cambridge University Press, 2009.

www.ingramcontent.com/pod-product-compliance
Lightning Source LLC
Chambersburg PA
CBHW040752220326
41597CB00029BA/4737